CHALLENGES AND STRATEGIES FOR DRYLAND AGRICULTURE

CHALLENGES AND STRATEGIES FOR DRYLAND AGRICULTURE

Dr. Parmeshwar Singh

RANDOM PUBLICATIONS

NEW DELHI - 110 002 (INDIA)

Challenges and Strategies for Dryland Agriculture

ISBN 978-93-51116-78-3

Published in 2015 in India by

RANDOM PUBLICATIONS

Reprint, 2017

4376-A/4B, Gali Murari Lal, Ansari Road

New Delhi-110 002

Phone: +9111-43580356, 23289044

E-mail: randomexports@gmail.com; sales@randompublications.com; info@randompublications.com

Type Setting by: Friends Media, Delhi-110089

Digitally Printed at: Replika Press Pvt. Ltd.

Preface

Dryland farming is an agricultural technique for non-irrigated cultivation of drylands. The choice of crop is influenced by the timing of the predominant rainfall in relation to the seasons. For example, winter wheat is more suited to regions with higher winter rainfall while areas with summer wet seasons may be more suited to summer growing crops such as sorghum, sunflowers or cotton. Dryland farming has evolved as a set of techniques and management practices used by farmers to continually adapt to the presence or lack of moisture in a given crop cycle. In marginal regions, a farmer should be financially able to survive occasional crop failures, perhaps for several years in succession. Survival as a dryland farmer requires careful husbandry of the moisture available for the crop and aggressive management of expenses to minimize losses in poor years. Dryland farming involves the constant assessing of the amount of moisture present or lacking for any given crop cycle and planning accordingly. Dryland farmers know that to financially succeed they have to be aggressive during the good years in order to offset the bad years.

Dryland farming is uniquely dependent on natural rainfall, which can leave the ground vulnerable to dust storms, particularly if poor farming techniques are used or if the storms strike at a particularly vulnerable time. The fact that a fallow period must be included in the crop rotation means that fields cannot always be protected by a cover crop, which might otherwise offer protection against erosion. Since healthy topsoil is critical to sustainable dryland agriculture, its preservation is generally considered the most important long-term goal of a dryland farming operation. Erosion control techniques such as windbreaks, reduced tillage or no-till, spreading straw (or other mulch on particularly susceptible ground), and strip farming are used to minimize topsoil loss. Dryland farming is practiced in regions inherently marginal for non-irrigated agriculture. Because of this, there is an increased risk of crop failure and poor yields which may occur in a dry year (regardless of money or effort expended). Dryland farmers must evaluate the potential yield of a crop constantly throughout the growing season and be prepared to decrease inputs to the crop such as fertilizer and weed control if it appears that it is likely to have a poor yield due to insufficient moisture. Conversely, in years when moisture is abundant, farmers may increase their input efforts and budget to maximize yields and to offset poor harvests. Dryland agriculture emerges as the biggest drag on the growth of the economy. Indeed, the dominant strand of thinking among our policy-makers treats the drylands as a hopelessly lost bet. Facing the challenge of the drylands is no longer a matter of choice. It is an imperative if we are to meet the goal of national food security in the coming years. Even in the most optimistic scenario of further irrigation development in India, nearly 40% of national demand for food in 2020 will have to be met through increasing the productivity of rainfed dryland agriculture.

This book will be useful resource guide for teachers, agricultural scientists, students and people engaged in NGO's associated with soil, water conservation and agriculture development.

I thank all members of my team who have helped in the preparation of the book. My special thanks go to "Random Publications" who have published the book.

— Dr. Parmeshwar Singh

Contents

Chapter 1

Agro-climatic Zones and Crop Classification

Meteorological conditions before, during and after application of agrochemicals are very important for both efficacy and crop safety. Also, the speed of growth and development of plants and crops, and of pests, is strongly influenced by meteorological parameters. Although micrometeorological circumstances can differ within small areas and although micrometeorological data is recorded before, during and after application of products in efficacy trials, it is better, when comparing areas, to use data on climate, because: "Climate is the synthesis of weather events over the whole of a period statistically long enough to establish its statistical ensemble properties (mean value, variation, probabilities of extreme events, etc.) and is largely independent of any instantaneous events."

In order to assess the importance of climate at different localities for any kind of trials, it is necessary to establish what relationships exist between the growth of specific crops and their pests and the values of certain climatic elements. Climatic elements of high importance for the efficacy and phytotoxicity of plant protection products, for the germination and growth of weeds and for rate of development of other pests are: temperature (mean, minimum, maximum), relative humidity, precipitation, radiation and wind (direction and velocity).

Most crops show distinct differences, within their area of distribution, in many of their properties, for example in phenological data, growth, yield and quality. The chief cause of such differences is to be sought in variations in climate. Even fairly small variations in a single climatic element can exert an obvious influence, for example in relation to cultivars of a given plant.

These differences need only occur during a few growth controlling periods, and plants are affected not only by differences in the single climatic elements but also by various combinations occurring simultaneously or in succession. For this reason, the usual division of Europe into a few areas with distinct values of the main climatic elements is inadequate from the point of view of agronomy. By taking into account all the climatic elements and making a more detailed classification of their respective values, it is possible to divide Europe into numerous subareas, each with its narrowly differentiated 'general type of agroclimate'.

Variety of Climates in Europe

Europe extends over 30 degrees of latitude in a north–south direction, which means that incident sunshine releases very different quantities of energy in different areas of Europe, particular during certain seasons. The very broken coastline in the south, west and north also plays an important part in determining the distribution of temperature and rainfall from west to east. All climate maps shows this multiplicity. In relation to crops, the main points of interest are: rainfall (during the growing period), temperature (mostly only temperature in the growing period, but for some crops, temperature throughout the year), frost-free periods and humidity.

Rainfall

Average annual rainfall is abundant nearly everywhere in Europe and in some areas of the Middle East. However, in the mountains, it is much heavier than in the surrounding lowlands (Walter, 1969). Prevailing winds over most of the area are westerly all the year, but they are stronger and more constant in the northwest. The maritime precipitation furnished by these prevailing moist westerly winds is blocked by coastal mountains in several places, but this only affects the climate locally and does not act as a complete barrier to the precipitation.

The east–west European system of mountains from Spain to the Balkans is, however, an important climatic divide in that it acts as a barrier to the Mediterranean climate from the south, as well as to the maritime climate from the west and northwest. The broad pattern of central Europe is one of transition from the oceanic northwest of Europe to continental Russia. From west to east, precipitation falls gradually (up to 40% reduction), except where highlands and mountains intervene.

Temperature and Frost-free Period

Temperature in Europe varies greatly along any parallel of latitude, owing to prevailing winds and variation in altitude. Average temperature decreases as the altitude increases and the rate of decrease is about the same everywhere. It falls about 1 °C for every 100 m of elevation. In northern and northwestern Europe, and in most of central Europe, the prevailing winds are from the west, and blow straight from the warm ocean, which is a more important source of heat to western Europe in winter than the direct rays of the sun. These winds are the cause of the mild, moist and cloudy climate of western and northern Europe.

A notable characteristic of the European climate is its lack of extreme variations in a particular area. Along most of the west coast of the continent, the difference between the mean temperature of the warmest and coldest months is less than 6.5°C. During the winter months, the centre of Europe is cold, but the northwestern part has a relatively high temperature because of the warm air currents coming in from the ocean. The coldest part of Europe in January is in the northeastern part of European Russia and the North of Scandinavia, while the warmest areas in winter are the Mediterranean islands and the tips of Mediterranean peninsulas. In some northern regions of Europe and Asia, the amount of radiation is a limiting factor in the growing season. The term 'frost-free season' is generally used to designate the number of days without killing frost in any region. It is determined strictly by the prevailing temperature conditions, and only expresses what may be called the 'thermal growing season' of a large area. It does not take into account local environmental factors, such as lack of moisture or poor soil. These may prevent a specific crop from being grown in an area, even where the season is satisfactorily frost-free.

Climate Comparability Definition of Zones

Many authors have published climate classification systems and maps showing the limits of the various climates, main climatic types, climatic types, climatic subtypes and transition climates. A review of this work was given by Essenwanger (2001). Classification of climate is a grouping of atmospheric conditions for locations which show similar climatic conditions (climate types) separated by defined boundaries applied to one or more meteorological elements (Essenwanger, 2001).

None of these classification systems is based on all climatic elements affecting the crop. A map was accordingly constructed showing all boundaries between areas of different climates within Europe. In this way, it should be possible to use all climatic parameters influencing crops and pests. In mountainous areas, the variety of climatic factors makes it impossible to give a general characterization for the entire zone.

Climate Diagrams

Walter & Lieth (1960) produced climate diagrams from many meteorological stations throughout the world. These are brief summaries of average climate variables and seasonal variation.

The diagrams display monthly averages for temperature and precipitation and the seasonal patterns over a year. They also give an idea of the duration of cold/warm or wet/dry seasons and the extent of their fluctuations. Since all the diagrams are plotted on the same scale, it is possible to compare moisture, temperature and other environmental conditions in widely separated parts of the world.

The boundaries between the 'general agro-climate' subareas hardly ever represent an abrupt transition from one climatic element to another. Besides, these classification systems are not based on all the climatic elements which affect crops, and in most cases they make no contribution to the solution of agricultural problems. Hence the present division into 'general agro-climate' subareas does not represent any new grouping but is a 'projection' on a single map of the divisions as conceived by Thran & Broekhuizen (1965).

Climate Classification System

Köppen & Geiger (1936) made a climate classification system based on a quantitative system of temperature and amount of precipitation. The Köppen system distinguishes six major types of regional climate (A, B, C, D, E, H), which are further divided into subcategories. The subcategories BW and BS take vegetation into account, and a further subdistinction between 'k' and 'h' distinguishes hot (h) from cool (k) climates.

According to the Köppen system, the European and Mediterranean region (EPPO region) belongs to regional climates C (moist climates with mild winters), D (moist climates with severe winters) and H (highland climates). Plants were used as meteorological instruments for measuring precipitation effectiveness and drainage to make the regional climates : Bhw, BWh, Cfa, Cfb, Csa, Csb, Cfa, Dfa, Dfb, Dfc.

One of the modifications of the system of Köppen & Geiger was made by Strahler owing to air-mass source regions and their zones of contact. Strahler's modification is based on latitudinal and coastal locations of places. He developed three groups of latitude climates, divided into 15 groups in all. This is a modification of a plant distribution system that first appeared in the USDA Yearbook, conforming to the principle that the presence and form of natural vegetation are mainly attributable to the qualities of the atmosphere. An easy way to recognize a climatic region is through its effect on the predominant groupings of plants that grow on the surface of the earth. An important modification of the basic principles of Köppen has been made by Thran & Broekhuizen (1965), by placing emphasis on the relevant time of year when evaluating climatic data. Thus, climate data for the winter season is not relevant for the pest problems of spring-sown crops. Only when the winter period is of significant importance to weeds (overwintering, emergence), insect pests (diapause, mortality) or pathogens (overwintering bodies) in relation to the autumn-sown crop or to a long-term crop, is it of major importance to look at the climate data for the full year.

Thran & Broekhuizen (1965) tried to create new agroclimatic subareas, based on the length of the period from spring to midsummer, and from midsummer to autumn. These 'phenological' periods, which provide a suitable basis for evaluating agricultural potential, refer to the period between the sowing of the earliest crop in spring to the sowing of the latest crop in the autumn. It gives some information on the time when spring begins (e.g. when the +5 °C mean daily temperature is exceeded and conditions remain frost-free) and the length of the vegetation period. Using the Köppen & Geiger classification system, Thran & Broekhuizen adjusted annual precipitation, annual temperature, summer temperature (warmest month), winter temperature (coldest month), precipitation and photoperiod in up to nine ranks. With the help of this ranking system, they developed 76 different agro-climatic subareas in Europe.

Proposed Zones

On the basis of all the information described above, and taking into account natural vegetation maps for Europe and Asia, we have been able to divide Europe, the Mediterranean area and the Middle East into a number of agro-climatic zones 3.

The Mediterranean zone comprises the countries or parts of countries around the Mediterranean sea, together with Jordan, Macedonia and Portugal. It is the region of 'Mediterranean' plant species, favoured by mild winter and warm summer temperatures with relatively wet winters and dry summers. This corresponds to the region with codes Csa and Csb according to Köppen & Geiger, code IV according to Walter & Lieth, and codes 33–36, 458, 65–75, V–Y according to Thran & Broekhuizen.

The Maritime zone is the zone north of the line from the coastal zone of southwest France, through Lyon (France), to the south border of Switzerland and Austria, west of the border between Austria and Hungary, west of the border between Czech Republic and Slovakia, west of the river Oder (between Poland and Germany). This zone also includes Ireland, Sweden and the United Kingdom. It is the region of 'Atlantic' plant species, which grow in moderately cool or cold winters and fairly mild summer temperatures, with relatively wet winters and wet to occasionally dry summers. This corresponds to the region with codes Cfa and Cfb according to Köppen & Geiger, code VI according to Walter & Lieth, and codes 3, 6, 7, 8, 14b, 17, 19–25, 27–33, 39–45 according to Thran & Broekhuizen.

The *North-east zone* includes the countries and the regions east of the river Oder (between Poland and Germany), north of the border between Czech Republic and Poland, west of the border between Poland and Ukraine, north of the border between Ukraine and Belarus, Russia north of 50°latitude. It is the region of 'Continental' plant species that grow in cold relatively wet winters and mild dry summers. This corresponds to the region with codes Dfb and Dfc according to Köppen & Geiger, code VIII according to Walter & Lieth, and codes 1, 4, 5, 14, 18, 60, A, B, E, F, G, I, M, H according to Thran & Broekhuizen.

The *Central zone*is made up of Bosnia-Herzegovina, Bulgaria, Croatia, Hungary, Moldova, Romania, Russia south of 50° latitude, Slovakia, Slovenia, Serbia and Montenegro, Turkey, Ukraine, except the Mediterranean coastal zones. It is the region of 'Continental' plant species that grow in cold relatively dry winters and warm dry to occasionally wet summers. This corresponds to the region with codes Cfb, Dfb and Bsk according to Köppen & Geiger, code VII according to Walter & Lieth, and codes 60–64, P, R, S, N, K, L according to Thran & Broekhuizen. This zonation is proposed for the comparability of efficacy evaluation data. A similar analysis can be made for data on residues of plant protection products, and results in only two zones (Northern Europe and Mediterranean).

Other Relevant Factors

Climate is only one of the factors to be considered in establishing the relevance of data on the evaluation of plant protection products from one part of the EPPO region to another. Some of the other factors that may need to be considered are presented below. It is not possible to provide a prescriptive list, as this depends on the individual circumstances of use, mode of action, physical and chemical properties of the product, etc. Instead, the case for comparability should focus on factors relevant for performance and crop safety, and for the biology and incidence of the target. Extrapolation between regions may still be valid when the factors differ significantly, provided that the situation, in terms of efficacy and crop safety of the plant protection product, can be considered to be at least as challenging in the area of 'recognition' as in the area of registered use.

Edaphic Conditions

For soil-applied products, it is important to compare soil types, organic matter content and pH. It may also be important to take into account soil moisture content/deficit over the duration of the trial.

Agronomic Conditions

Cultural conditions and agronomy which may need to be considered include soil cultivation, application methods, cultivars, fertilizer regime, times of planting and harvest.

Differences in Target Pest Pressure

Certain aspects related to the target pests may give rise to a more or less harsh test of the efficacy and crop safety of the plant protection product. These could include: differences in epidemiology or population dynamics of the pests, different races of the target pest, resistance of pest populations.

Ecological Divisions of India-Agricultural Meteorology

A systematic appraisal of the soil and climatic resources is a pre-requisite for formulating effective land use plan for various regions of our country. Mapping of the various agro-ecological regions will help in identifying suitable cropping patterns for a particular region. There are several classifications of agro-climatic regions and soils proposed by several agencies. The reader is advised to study the earlier articless on the related topics Land Capability Classification and Agro-climatic Divisions of India. This article is on the agro-ecological regions of India; but there may be several things similar to the article on Agro-climatic Divisions of India. We must keep in mind the distinction between agroclimatic divisions and agro-ecological regions even though there may be similarities.

Indian sub-continent exhibits a variety of landscapes and climatic conditions which are noticeable in the types of soils and vegetation. Needless to say that there exists a significant relationship between climate, soils, type of land and vegetation. While preparing for land use plan or a cropping pattern or agricultural or animal husbandry system we should keep In mind the combine effect of climate, soil, land formation, topography and vegetation of the place.

In the past, several attempts have been made to understand and classify the agro-climatic regions as well as the agro-ecological regions of our country. In the year 1954, Carter divided India into six climatic regions, ranging from arid to perhumid, based on the criteria of Thornthwhite system of climatic classification. Murthy and Pandey (1978) on the basis of physiography, climate (rainfall and potential water surplus/deficit), soils and agricultural regions identified eight agroecological regions. Though this was a good beginning this system of classification suffered from some limitations. Based on the data of 160 meteorological stations in the country and using the concept of moisture adequacy index Subramaniam (1983) proposed 29 agro-ecological zones with the possible 36 combinations of moisture adequacy index and dominant soil groups following the pattern of FAO/UNESCO. Sehgal et al.(1987) prepared a computerized bio-climatic map of North West India, based on the criteria of dry month (the month having the real moisture deficit of 40% or more of the potential evapo-transpiration-PET). Krishnaft (1988) delineated 40 soil-climatic zones based on major soil types and moisture index.

Planning Commission as part of the mid-term appraisal of the planning targets of VII Plan (1985-1990) divided the country into 15 broad agro-climatic zones based on physiography and climate. They are the following: 1. Western Himalayan Region, 2. Eastern Himalayan I Region, 3. Lower Gangetic Plains Region, 4. Middle Gangetic Plains Region, 5. Upper Gangetic Plains Region, 6. Trans-Gangetic Plains Region, 7. Eastern Plateau and Hills Region, I 8. Central Plateau and Hills Region, 9. Western Plateau and Hills Region, 10. Southern Plateau and Hills Region, 11. East Coast Plains and Hills Region, 12. West Coast plains and Hills Region, 13. Gujarat Plains and Hills Region, 14. Western Dry Region and 15. The Island Region.

The state agricultural universities were entrusted with the task of dividing these agroclimatic zones into state-wise sub-agro-climatic zones under the National Agricultural Research Project (NARP). Accordingly 127 sub zones were identified based on the rainfall, existing cropping pattern and administrative units.

The government of India was not satisfied with the classifications done so far. It further entrusted the National Bureau of Soil Survey and Land Use Planning (NBSS&LUP) of Indian Council of Agricultural Research to prepare an agro-ecological region map of the country based on the parameters (rainfall, temperature, vegetation, potential evapo-transpiration) which form the-Bioclimate-and parameters (rainfall, potential evapotranspiration, soil storage) which constitutes the-Length of the Growing Period-(LGP) and parameters (soils and physiography) which form the-Soil or Land Scape. This article describes this classification.

Basis of Classification

The parameters taken for the classification agro-ecological regions are the characteristics of: physiographical features, soil characteristics, bio-climatic types and length of the growing period. These are explained one by one briefly.

Physiographic Features

The major physiographic regions in our country are the Mountainous region, Indo-Gangetic Alluvial Plains, Peninsular or Deccan Plateau and Coastal Plains. These are further divided into 19 broad and basic agro-ecological regions. Each region is represented by a capital letter of the English alpha-bets as given here.

Westem Himalayas

This includes Ladakh plateau, Kashmir Himalayas, Punjab Himalayas and Kumaun Himalayas of Uttar Pradesh and is symbolized by-A.

Central Himalayas

This includes mainly Nepal Himalayas, and foot-hills of V.P. Himalayas and is symbolized by-B.

Climate and its Influence on Crop Production

Climate change and agriculture are interrelated processes, both of which take place on a global scale. Global warming is projected to have significant impacts on conditions affecting agriculture, including temperature, carbon dioxide, glacial run-off, precipitation and the interaction of these elements. These conditions determine the carrying capacity of the biosphere to produce enough food for the human population and domesticated animals. The overall effect of climate change on agriculture will depend on the balance of these effects. Assessment of the effects of global climate changes on agriculture might help to properly anticipate and adapt farming to maximize agricultural production.

At the same time, agriculture has been shown to produce significant effects on climate change, primarily through the production and release of greenhouse gases such as carbon dioxide, methane, and nitrous oxide, but also by altering the Earth's land cover, which can change its ability to absorb or reflect heat and light, thus contributing to radiative forcing. Land use change such as deforestation and desertification, together with use of fossil fuels, are the major anthropogenic sources of carbon dioxide; agriculture itself is the major contributor to increasing methane and nitrous oxide concentrations in earth's atmosphere.

Impact of Climate Change on Agriculture

Despite technological advances, such as improved varieties, genetically modified organisms, and irrigation systems, weather is still a key factor in agricultural productivity, as well as soil properties and natural communities. The effect of climate on agriculture is related to variabilities in local climates rather than in global climate patterns. The Earth's average surface temperature has increased by 1 degree F in just over the last century. Consequently, agronomists consider any assessment has to be individually consider each local area. On the other hand, agricultural trade has grown in recent years, and now provides significant amounts of food, on a national level to major importing countries, as well as comfortable income to exporting ones. The international aspect of trade and security in terms of food implies the need to also consider the effects of climate change on a global scale.

A study published in *Science* suggest that, due to climate change, "southern Africa could lose more than 30% of its main crop, maize, by 2030. In South Asia losses of many regional staples, such as rice, millet and maize could top 10%". The 2001 IPCC Third Assessment Report concluded that the poorest countries would be hardest hit, with reductions in crop yields in most tropical and sub-tropical regions due to decreased water availability, and new or changed insect pest incidence. In Africa and Latin America many rainfed crops are near their maximum temperature tolerance, so that yields are likely to fall sharply for even small climate changes; falls in agricultural productivity of up to 30% over the 21st century are projected. Marine life and the fishing industry will also be severely affected in some places.

Climate change induced by increasing greenhouse gases is likely to affect crops differently from region to region. For example, average crop yield is expected to drop down to 50% in Pakistan according to the UKMO scenario whereas corn production in Europe is expected to grow up to 25% in optimum hydrologic conditions.

More favourable effects on yield tend to depend to a large extent on realization of the potentially beneficial effects of carbon dioxide on crop growth and increase of efficiency in water use. Decrease in potential yields is likely to be caused by shortening of the growing period, decrease in water availability and poor vernalization.

In the long run, the climatic change could affect agriculture in several ways :

- *productivity*, in terms of quantity and quality of crops
- *agricultural practices*, through changes of water use (irrigation) and agricultural inputs such as herbicides, insecticides and fertilizers
- *environmental effects*, in particular in relation of frequency and intensity of soil drainage (leading to nitrogen leaching), soil erosion, reduction of crop diversity
- *rural space*, through the loss and gain of cultivated lands, land speculation, land renunciation, and hydraulic amenities.
- *adaptation*, organisms may become more or less competitive, as well as humans may develop urgency to develop more competitive organisms, such as flood resistant or salt resistant varieties of rice.

They are large uncertainties to uncover, particularly because there is lack of information on many specific local regions, and include the uncertainties on magnitude of climate change, the effects of technological changes on productivity, global food demands, and the numerous possibilities of adaptation. Most agronomists believe that agricultural production will be mostly affected by the severity and pace of climate change, not so much by gradual trends in climate. If change is gradual, there may be enough time for biota adjustment. Rapid climate change, however, could harm agriculture in many countries, especially those that are already suffering from rather poor soil and climate conditions, because there is less time for optimum natural selection and adaption.

Projections

The literature on key vulnerabilities to climate change. With low to medium confidence, they concluded that for about a 1 to 3°C global mean temperature increase (by 2100, relative to the 1990-2000 average level) there would be productivity decreases for some cereals in low latitudes, and productivity increases in high latitudes. With medium confidence, global production potential was predicted to:

- increase up to around 3°C,
- very likely decrease above about 3 to 4°C.

Most of the studies on global agriculture assessed by Schneider *et al.*. (2007:790) had not incorporated a number of critical factors, including changes in extreme events, or the spread of pests and diseases. Studies had also not considered the development of specific practices or technologies to aid adaptation.

Regional

- Africa:
  * Africa's geography makes it particularly vulnerable to climate change, and seventy per cent of the population rely on rain-fed agriculture for their livelihoods. Tanzania's official report on climate change suggests that the areas that usually get two rainfalls in the year will probably get more, and those that get only one rainy season will get far less. The net result is expected to be that 33% less maize—the country's staple crop—will be grown. Alongside other factors, regional climate change-in particular, reduced precipitation-is thought to have contributed to the conflict in Darfur. The combination of decades of drought, desertification and overpopulation are among the causes of the conflict, because the Baggara Arab nomads searching for water have to take their livestock further south, to land mainly occupied by farming peoples.
  * With high confidence, IPCC (2007:13) concluded that climate variability and change would severely compromise agricultural production and access to food.
- Asia: With medium confidence, IPCC (2007:13) projected that by the mid-21st century, in East and Southeast Asia, crop yields could increase up to 20%, while in Central and South Asia, yields could decrease by up to 30%. Taken together, the risk of hunger was projected to remain very high in several developing countries.
- Australia and New Zealand: Hennessy *et al.*. (2007:509) assessed the literature for this region. They concluded that without further adaptation to climate change, projected impacts would likely be substantial: By 2030, production from agriculture and forestry was projected to decline over much of southern and eastern Australia, and over parts of eastern New Zealand; In New Zealand, initial benefits were projected close to major rivers and in western and southern areas. Hennessy *et al.*. (2007:509) placed high confidence in these projections.
- Europe: With high confidence, IPCC (2007:14) projected that in Southern Europe, climate change would reduce crop

productivity. In Central and Eastern Europe, forest productivity was expected to decline. In Northern Europe, the initial effect of climate change was projected to increase crop yields.

- Latin America: With high confidence, IPCC (2007:14) projected that in drier areas of Latin America, productivity of some important crops would decrease and livestock productivity decline, with adverse consequences for food security. In temperate zones, soybean yields were projected to increase.
- North America:
  * According to a paper by Deschenes and Greenstone (2006), predicted increases in temperature and precipitation will have virtually no effect on the most important crops in the US.
  * With high confidence, IPCC (2007:14-15) projected that over the first few decades of this century, moderate climate change would increase aggregate yields of rain-fed agriculture by 5–20%, but with important variability among regions. Major challenges were projected for crops that are near the warm end of their suitable range or which depend on highly utilized water resources.
- Polar regions (Arctic and Antarctic):
  * For the *Guardian* newspaper, Brown (2005) reported on how climate change had affected agriculture in Iceland. Rising temperatures had made the widespread sowing of barley possible, which had been untenable twenty years ago. Some of the warming was due to a local (possibly temporary) effect via ocean currents from the Caribbean, which had also affected fish stocks.
  * Anisimov *et al.*. (2007:655) assessed the literature for this region. With medium confidence, they concluded that the benefits of a less severe climate were dependent on local conditions. One of these benefits was judged to be increased agricultural and forestry opportunities.
- Small islands: In a literature assessment, Mimura *et al.*. (2007:689) concluded, with high confidence, that subsistence and commercial agriculture would very likely be adversely affected by climate change.

Shortage in Grain Production

Between 1996 and 2003, grain production has stabilized slightly over 1800 millions of tons. In 2000, 2001, 2002 and 2003, grain stocks have been dropping, resulting in a global grain harvest that was short of consumption by 93 millions of tons in 2003.

The Earth's average temperature has been rising since the late 1970s, with nine of the 10 warmest years on record occurring since 1995. In 2002, India and the United States suffered sharp harvest reductions because of record temperatures and drought. In 2003 Europe suffered very low rainfall throughout spring and summer, and a record level of heat damaged most crops from the United Kingdom and France in the Western Europe through Ukraine in the East. Bread prices have been rising in several countries in the region.

Poverty Impacts

Researchers at the Overseas Development Institute (ODI) have investigated the potential impacts climate change could have on agriculture, and how this would affect attempts at alleviating poverty in the developing world.

They argued that the effects from moderate climate change are likely to be mixed for developing countries. However, the vulnerability of the poor in developing countries to short term impacts from climate change, notably the increased frequency and severity of adverse weather events is likely to have a negative impact. This, they say, should be taken into account when defining agricultural policy.

Crop Development Models

Models for climate behaviour are frequently inconclusive. In order to further study effects of global warming on agriculture, other types of models, such as *crop development models*, *yield prediction*, quantities of *water or fertilizer consumed*, can be used. Such models condense the knowledge accumulated of the climate, soil, and effects observed of the results of various agricultural practices. They thus could make it possible to test strategies of adaptation to modifications of the environment. Because these models are necessarily simplifying natural conditions (often based on the assumption that weeds, disease and insect pests are controlled), it is not clear whether the results they give will have an *in-field* reality. However, some results are partly validated with an increasing number of experimental results.

Other models, such as *insect and disease development* models based on climate projections are also used (for example simulation of aphid reproduction or septoria (cereal fungal disease) development).

Scenarios are used in order to estimate climate changes effects on crop development and yield. Each scenario is defined as a set of meteorological variables, based on generally accepted projections. For example, many models are running simulations based on doubled carbon dioxide projections, temperatures raise ranging from 1°C up to 5°C, and with rainfall levels an increase or decrease of 20%. Other parameters may include humidity, wind, and solar activity. Scenarios of crop models are testing farm-level adaptation, such as sowing date shift, climate adapted species (vernalisation need, heat and cold resistance), irrigation and fertilizer adaptation, resistance to disease. Most developed models are about wheat, maize, rice and soybean.

Temperature Potential Effect on Growing Period

Duration of crop growth cycles are above all, related to temperature. An increase in temperature will speed up development. In the case of an annual crop, the duration between sowing and harvesting will shorten (for example, the duration in order to harvest corn could shorten between one and four weeks). The shortening of such a cycle could have an adverse effect on productivity because senescence would occur sooner.

Effect of Elevated Carbon Dioxide on Crops

Carbon dioxide is essential to plant growth. Rising CO_2 concentration in the atmosphere can have both positive and negative consequences. Increased CO_2 is expected to have positive physiological effects by increasing the rate of photosynthesis. Currently, the amount of carbon dioxide in the atmosphere is 380 parts per million. In comparison, the amount of oxygen is 210,000 ppm. This means that often plants may be starved of carbon dioxide, due to the enzyme that fixes CO_2, rubisco also fixes oxygen in the process of photorespiration. The effects of an increase in carbon dioxide would be higher on C3 crops (such as wheat) than on C4 crops (such as maize), because the former is more susceptible to carbon dioxide shortage. Studies have shown that increased CO_2 leads to fewer stomata developing on plants which leads to reduced water usage. Under optimum conditions of temperature and humidity, the yield increase could reach 36%, if the levels of carbon dioxide are doubled.

Further, few studies have looked at the impact of elevated carbon dioxide concentrations on whole farming systems. Most models study the relationship between CO_2 and productivity in isolation from other factors associated with climate change, such as an increased frequency of extreme weather events, seasonal shifts, and so on. In 2005, the Royal Society in London concluded that the purported benefits of elevated carbon dioxide concentrations are "likely to be far lower than previously estimated" when factors such as increasing ground-level ozone are taken into account."

Effect on Quality

According to the IPCC's TAR, "The importance of climate change impacts on grain and forage quality emerges from new research. For rice, the amylose content of the grain—a major determinant of cooking quality—is increased under elevated CO_2". Cooked rice grain from plants grown in high-CO_2 environments would be firmer than that from today's plants. However, concentrations of iron and zinc, which are important for human nutrition, would be lower. Moreover, the protein content of the grain decreases under combined increases of temperature and CO_2." Studies using FACE have shown that increases in CO_2 lead to decreased concentrations of micronutrients in crop plants. This may have knock-on effects on other parts of ecosystems as herbivores will need to eat more food to gain the same amount of protein.

Studies have shown that higher CO_2 levels lead to reduced plant uptake of nitrogen (and a smaller number showing the same for trace elements such as zinc) resulting in crops with lower nutritional value. This would primarily impact on populations in poorer countries less able to compensate by eating more food, more varied diets, or possibly taking supplements. Reduced nitrogen content in grazing plants has also been shown to reduce animal productivity in sheep, which depend on microbes in their gut to digest plants, which in turn depend on nitrogen intake.

Agricultural Surfaces and Climate Changes

Climate change may increase the amount of arable land in high-latitude region by reduction of the amount of frozen lands. A 2005 study reports that temperature in siberia has increased three degree Celsius in average since 1960 (much more than the rest of the world). However, reports about the impact of global warming on Russian agriculture indicate conflicting probable effects : while they expect a northward extension of farmable lands, they also warn of possible productivity losses and increased risk of drought. Sea levels are expected to get up to one meter higher by 2100, though this projection is disputed. A rise in the sea level would result in an agricultural land loss, in particular in areas such as South East Asia. Erosion, submergence of shorelines, salinity of the water table due to the increased sea levels, could mainly affect agriculture through inundation of low-lying lands.

Low lying areas such as Bangladesh, India and Vietnam will experience major loss of rice crop if sea levels are expected to rise by the end of the century. Vietnam for example relies heavily on its southern tip, where the Mekong Delta lies, for rice planting. Any rise in sea level of no more than a meter will drown several km^2. of rice paddies, rendering Vietnam incapable of producing its main staple and export of rice.

Erosion and Fertility

The warmer atmospheric temperatures observed over the past decades are expected to lead to a more vigorous hydrological cycle, including more extreme rainfall events. Erosion and soil degradation is more likely to occur. Soil fertility would also be affected by global warming. However, because the ratio of carbon to nitrogen is a constant, a doubling of carbon is likely to imply a higher storage of nitrogen in soils as nitrates, thus providing higher fertilizing elements for plants, providing better yields. The average needs for nitrogen could decrease, and give the opportunity of changing often costly fertilisation strategies.

Due to the extremes of climate that would result, the increase in precipitations would probably result in greater risks of erosion, whilst at the same time providing soil with better hydration, according to the intensity of the rain. The possible evolution of the organic matter in the soil is a highly contested issue: while the increase in the temperature would induce a greater rate in the production of minerals, lessening the soil organic matter content, the atmospheric CO_2 concentration would tend to increase it.

The warmer atmospheric temperatures observed over the past decades are expected to lead to a more vigorous hydrological cycle, including more extreme rainfall events. In 1998 Karl and Knight reported that from 1910 to 1996 total precipitation over the contiguous U.S. increased, and that 53% of the increase came from the upper 10% of precipitation events (the most intense precipitation). The percent of precipitation coming from days of precipitation in excess of 50 mm has also increased significantly.

Studies on soil erosion suggest that increased rainfall amounts and intensities will lead to greater rates of erosion. Thus, if rainfall amounts and intensities increase in many parts of the world as expected, erosion will also increase, unless amelioration measures are taken. Soil erosion rates are expected to change in response to changes in climate for a variety of reasons.

The most direct is the change in the erosive power of rainfall. Other reasons include: a) changes in plant canopy caused by shifts in plant biomass production associated with moisture regime; b) changes in litter cover on the ground caused by changes in both plant residue decomposition rates driven by temperature and moisture dependent soil microbial activity as well as plant biomass production rates; c) changes in soil moisture due to shifting precipitation regimes and evapotranspiration rates, which changes infiltration and runoff ratios; d) soil erodibility changes due to decrease in soil organic matter concentrations in soils that lead to a soil structure that is more susceptible to erosion and increased runoff due to increased soil surface sealing and crusting; e) a shift of winter precipitation from non-erosive snow to erosive rainfall due to increasing winter temperatures; f) melting of permafrost, which induces an erodible soil state from a previously non-erodible one; and g) shifts in land use made necessary to accommodate new climatic regimes.

Effects of Global Climate Change on Pests, Diseases and Weeds

A very important point to consider is that weeds would undergo the same acceleration of cycle as cultivated crops, and would also benefit from carbonaceous fertilization. Since most weeds are C3 plants, they are likely to compete even more than now against C4 crops such as tomatoes. However, on the other hand, some results make it possible to think that weedkillers could gain in effectiveness with the temperature increase.

Global warming would cause an increase in rainfall in some areas, which would lead to an increase of atmospheric humidity and the duration of the wet seasons. Combined with higher temperatures, these could favour the development of fungal diseases. Similarly, because of higher temperatures and humidity, there could be an increased pressure from insects and disease vectors.

Glacier Retreat and Disappearance

The continued retreat of glaciers will have a number of different quantitative impacts. In areas that are heavily dependent on water runoff from glaciers that melt during the warmer summer months, a continuation of the current retreat will eventually deplete the glacial ice and substantially reduce or eliminate runoff. A reduction in runoff will affect the ability to irrigate crops and will reduce summer stream flows necessary to keep dams and reservoirs replenished.

Approximately 2.4 billion people live in the drainage basin of the Himalayan rivers. India, China, Pakistan, Afghanistan, Bangladesh, Nepal and Myanmar could experience floods followed by severe droughts in coming decades. In India alone, the Ganges provides water for drinking and farming for more than 500 million people. The west coast of North America, which gets much of its water from glaciers in mountain ranges such as the Rocky Mountains and Sierra Nevada, also would be affected.

Ozone and UV-B

Some scientists think agriculture could be affected by any decrease in stratospheric ozone, which could increase biologically dangerous ultraviolet radiation B. Excess ultraviolet radiation B can directly effect plant physiology and cause massive amounts of mutations, and indirectly through changed pollinator behaviour, though such changes are simple to quantify. However, it has not yet been ascertained whether an increase in greenhouse gases would decrease stratospheric ozone levels. In addition, a possible effect of rising temperatures is significantly higher levels of ground-level ozone, which would substantially lower yields.

ENSO Effects on Agriculture

ENSO (El Niño Southern Oscillation) will affect monsoon patterns more intensely in the future as climate change warms up the ocean's water. Crops that lie on the equatorial belt or under the tropical Walker circulation, such as rice, will be affected by varying monsoon patterns and more unpredictable weather. Scheduled planting and harvesting based on weather patterns will become less effective.

Areas such as Indonesia where the main crop consists of rice will be more vulnerable to the increased intensity of ENSO effects in the future of climate change. University of Washington professor, David Battisti, researched the effects of future ENSO patterns on the Indonesian rice agriculture using [IPCC]'s 2007 annual report and 20 different logistical models mapping out climate factors such as wind pressure, sea-level, and humidity, and found that rice harvest will experience a decrease in yield. Bali and Java, which holds 55% of the rice yields in Indonesia, will be likely to experience 9-10% probably of delayed monsoon patterns, which prolongs the hungry season. Normal planting of rice crops begin in October and harevest by January. However, as climate change affects ENSO and consequently delays planting, harvesting will be late and in drier conditions, resulting in less potential yields.

Environmental Issues with Agriculture

There are numerous environmental issues with the various practices of agriculture.

Climate Change

Climate change and agriculture are interrelated processes, both of which take place on a global scale. Global warming is projected to have significant impacts on conditions affecting agriculture, including temperature, precipitation and glacial run-off. These conditions determine the carrying capacity of the biosphere to produce enough food for the human population and domesticated animals. Rising carbon dioxide levels would also have effects, both detrimental and beneficial, on crop yields. The overall effect of climate change on agriculture will depend on the balance of these effects. Assessment of the effects of global climate changes on agriculture might help to properly anticipate and adapt farming to maximize agricultural production.

At the same time, agriculture has been shown to produce significant effects on climate change, primarily through the production and release of greenhouse gases such as carbon dioxide, methane, and nitrous oxide, but also by altering the Earth's land cover, which can change its ability to absorb or reflect heat and light, thus contributing to radiative forcing. Land use change such as deforestation and desertification, together with use of fossil fuels, are the major anthropogenic sources of carbon dioxide; agriculture itself is the major contributor to increasing methane and nitrous oxide concentrations in earth's atmosphere.

Deforestation

One of the causes of deforestation is to clear land for pasture or crops. According to British environmentalist Norman Myers, 5% of deforestation is due to cattle ranching, 19% due to over-heavy logging, 22% due to the growing sector of palm oil plantations, and 54% due to slash-and-burn farming.

In 2000 the United Nations Food and Agriculture Organization (FAO) found that "the role of population dynamics in a local setting may vary from decisive to negligible," and that deforestation can result from "a combination of population pressure and stagnating economic, social and technological conditions."

Genetic Engineering

Genetic engineering has caused controversies.

Seed contamination is problematic.

Intensive Farming

Intensive farming alters the environment in many ways. Some of the disadvantages of this method of farming include:

- Limits or destroys the natural habitat of most wildlife, and leads to soil erosion
- Use of fertilizers can alter the biology of rivers and lakes.
- Pesticides generally kill useful insects as well as those that destroy crops
- Generally not sustainable-often results in desertification or, in a worst case scenario, land that is so poisonous and eroded that nothing else will grow
- Requires large amounts of energy input to produce, transport, and apply chemical fertilizers/pesticides
- Use of chemicals on fields creates run-off, excess runs off into rivers and lakes causing pollution
- Use of pesticides have numerous negative health effects in workers who apply them, people that live nearby the area of application or downstream/downwind from it, and consumers who eat the pesticides which remain on their food

Irrigation

Irrigation can lead to a number of problems :

- Depletion of underground aquifers through overdrafting.
- Ground subsidence.
  - o Groundwater recharge-an ecological restoration, mitigation, and remediation technique.
- Underirrigation gives poor soil salinity control which leads to increased soil salinity with consequent build up of toxic salts on soil surface in areas with high evaporation. This requires either leaching to remove these salts and a method of drainage to carry the salts away.
- Overirrigation because of poor distribution uniformity or management wastes water, chemicals, and may lead to water pollution.
- Deep drainage (from over-irrigation) may result in rising water tables which in some instances will lead to problems of irrigation salinity requiring watertable control by some form of subsurface land drainage.
- Irrigation with saline or high-sodium water may damage soil structure owing to the formation of alkaline soil.
- Runoff causing surface water and groundwater-aquifer hydrologic cycle water pollution.
  - o Bioretention-an ecological restoration, mitigation, and remediation technique.

Pollutants

A wide range of agricultural chemicals are used and some become pollutants through use, misuse, or ignorance.

- Pesticide drift
    - soil contamination
    - groundwater and water pollution
    - air pollution *spray drift*
- Pesticides, especially those based on organochloride
- Pesticide residue in foods
- Pesticide toxicity to bees
    - List of crop plants pollinated by bees
    - Pollination management
- Bioremediation.

Waste

Plasticulture, the use of plastic materials in agriculture, raises problems around how to carry out the recycling of agricultural plastics.

Issues by Region

- Hedgerow removal in the United Kingdom.
- Soil salinisation, especially in Australia.
- Phosphate mining in Nauru
- Methane emissions from livestock in New Zealand.
- Some environmentalists attribute the hypoxic zone in the Gulf of Mexico as being encouraged by nitrogen fertilization of the algae bloom.

Sustainable Agriculture

The exponential population increase in recent decades has increased the practice of agricultural land conversion to meet demand for food which in turn has increased the effects on the environment. The global population is still increasing and will eventually stabilise, as some critics doubt that food production, due to lower yields from global warming, can support the global population. Organic farming is a multifaceted sustainable agriculture set of practices that can have a lower impact on the environment. Other specific methods include: permaculture; and biodynamic agriculture which incorporating a spiritual element.

- Category: Sustainable agriculture
- Biological pest control.

Chapter 2

Cold Arid Horticulture: An Overview

Indigenous knowledge in the Himalayan region is the inter-generational wisdom of local inhabitants to perform their livelihood operations in a most eco-friendly manner under remote, isolated and inaccessible conditions; characterized by harsh climate and limited survival options. Since this knowledge is transferred orally from one generation to the next, it is dynamic in dissemination and scientific in indigenous experimentation; receiving constant stimuli from outside. However, indigenous not only stands for ingrained intrinsic knowledge, but is also amenable to modifications based on latest technical know-how by local inhabitants through native means to suit their daily requirements. Therefore, to discard any indigenous knowledge on connotations of superstition, conservatism, primitivism etc. by modem science would only result into a failure of the developmental networks. Evidences have shown that the developmental projects which overtly rejected already acquired knowledge of the local inhabitants, have failed to achieve their targets.

Throughout Himalayan region, watershed resource use and productivity is based on crops, horticulture, pastures and forestry which is largely influenced by geographical and environment diversity prevailing in its different zones.

The wide variations in altitude and other agro-climatic parameters such as rainfall and temperature, broadly classify Himalayan region into four major agro-climatic zones. These include (1) the low hills and valleys near the plains, (2) the middle hills and valleys with sub-humid climate, (3) high mountains and valleys with temperate climate and (4) cold dry desert zone. In this document, ITK for upper Himalayas comprising of high mountains and valleys with temperate climate and cold dry deserts are described/presented. Before turning to ITK *per se,* it is pertinent to present a brief glimpse of the two regions under consideration. First, the cold desert region and then the temperate zone of the western Himalayas/Indian Upper Himalayas.

Cold Deserts in Western Himalayas

Spread over an approximate area of 74, 809 sq. km, the cold desert area in India covers 12 out of 131 desert blocks in India. Leh and Kargil districts of Ladakh in Jammu & Kashmir and Lahaul and Spiti along with some parts of Chamba and Kinnaur districts of Himachal Pradesh comprise this cold desert area. Similarity in their physiographic location and the consequent geomorphic processes still unfolding in the region lend to this entire region a largely similar texture.

Located in the interior of continents, away from any source of moisture, cold deserts manifest remarkable ecological variety and biological diversity. Their geographic remoteness and unfriendly climatic conditions greatly constrain economic growth and development. Environmental degradation, which is on increase, is an additional cause for concern.

Bio-physical Features

A rarified atmosphere, fast blowing winds -eroding the immature sandy soils, extreme variations in daily and seasonal temperatures along with scanty or no precipitation during spring and summer ensure short growing seasons (2-5 months) with exposure to harmful infra-red and ultraviolet radiations. Unharvested glacial melts, frozen soil moisture during early spring and low relative humidity during the growing season are some abiotic features. Sparce natural vegetation results from over-exploitation by a variety of agencies - e.g. grazing by both domestic and migratory animals, harvesting (of vegetation) to meet energy needs viz. fuelwood and dry fodder for winter besides demands from pharmaceutical agencies. Additionally, both migratory birds and rodents utilize dispersed seeds as a means of sustenance thereby jeopardizing natural regeneration.

Features such as high transpiration due to excessive heat (often causing mortality), inadequate photo-hours especially during winter, injury due to frost causes poor seed germination, poor plant growth, poor root formation, deformed canopy, reduced radial growth etc. and other physical signs/phenotypic manifestations which in turn affect the productive biomass production in the region.

Natural vegetation is overwhelmingly herbaceous - comprising of a few tree species and a few shrub species. *Juniperus wallichiana, J. communis, Caragana* spp., *Artemisea* spp., *Lonicera* spp., *Potentilla* spp., *Myricaria* spp., *Koleresia dutheii, Ephedra, Salix,* spp., *Juniperus* spp., *Rosa* spp., *Caragana* spp. *Rhododendron* spp., *Betula utilis* are found here. Additionally, manmade forests of poplars, willows, *Hippophae* spp. and *Myricaria* spp. can also be seen along river banks, rivulets and *nallahs.*

The herbaceous element is comprised of *Thymus, Medicago, Trifolium, Anemone, Potentila, Epilobium, Verbena, Allium, Aconitum, Delphenium, Aquilegia, Primula, Geranium, Polygonum* and *Cannabis.* This abundance of the herbaceous element, both in Laddakh and in the cold desert of Himachal, has been the mainstay of the traditional medicinal system prevalent in this region. The nature of flora along with man made interventions have ensured a land use pattern typical to this region.

Socio-economic Features

These features also place constraints on economic growth. Fundamentally these are re-enforced by the biophysical features discussed earlier. Society steeped in religion along with class/case hierarchies, absence of adequate means of livelihood, absence of women's organizations accompanied by rudimentary infrastructure - viz. poor communication network, inadequate developmental institutions/agencies, lack of educational/vocational facilities and inadequate outreach for relevant transfer of technology characterize cold desert society. The result, quite obviously, is that there remains a largely untapped and underdeveloped industrial potential even in the agriculture/horticulture sector that are the mainstay of communities inhabiting cold deserts.

In this region as a whole the initial land use pattern was purely agriculture. It has, however, changed over a period of time to agri-horti-silvi-pastoral. Apple, potato, walnut etc. are some important horticultural crops while poplars and willows have been popularised especially in Lahaul and Spiti and Laddakh to increase both the green cover and also to augment fodder and fuel needs.

The physiographic location of Laddakh enables this district to have only one cropping season - Kharif, which extends from March/April to October. Further depending upon altitude the growing season varies between two months (above 4000 in amsl) to five months (below 3000 in amsl). Millets followed by wheat are the most important cereal crops.

These are followed, in order of importance by fodder crops, barley and pulses. *Alfa-alfa* is the most popular fodder crop. Fruits and vegetables - the horticultural component, are a recent development. Since they yield high economic returns increasingly more and more area is being brought under crops such as apples, raisin grapes and apricots. Mustard, Pea, Lathyrus, millets and turnips are also grown. The central belt (3000-3500 in amsl) is very well suited to the production of vegetable seeds. Rotation in the cropping pattern is hardly practiced.

The fauna of this entire region is quite unique. Due to poor/ rudimentary communication facilities the yak has been the major animal for burden. Besides yaks, livestock comprise mainly sheep and goats.

While the major fauna species of cold desert in Jammu & Kashmir are snow leopard, ibex, snow cock, partridges, magpie etc. A large number of migratory birds visit the lakes and rivers. Pashmina goat, Changthangi sheep, yaks, donkeys and double humped camels are animals of economic importance. The ibex, bharal, brown bear, tibetan wolf, nayan, marmot, snow leopard, lynx, weasel, vole, snow cock, snow partridge, chukor, chough, raven etc. are found in the cold deserts of Himachal Pradesh. This uniqueness is being preserved through the establishment of two sanctuaries namely the Pin valley National Park in Spin and the Sechu Tuan Nala in Chamba.

As mentioned, yak and sheep dominate livestock composition. Additionally land extensive management and transhumance along with the barter system of converting livestock into other usable commodities are features that characterise animal husbandry practices. Chief among the many problems faced by the livestock sector are - insufficient supply of fodder, overgrazing right up to alpine meadows and difficulty in stall feeding in snow bound areas etc. The *Gaddis* - a migratory tribe of Bharmour, use goat milk to supplement their diet.

Temperate Zone in Western Himalayas

The temperate zone in western Himalayas corresponds to zone III drawn up on agro-ecological basis by the planning commission. It is physiographically a largely mountaneous tract (1800-2000 m amsl).

This zone is characterised by mountaneous tracts of varying altitudes, steep slopes etc. Steeper slopes receive higher intensity of radiation and are not conducive to the growth of vegetation. Conversely, vegetation abounds on the relatively gentler slopes. Moisture from snow-melt and rich organic matter generate dense vegetation of fruit and forest trees. Most of this zone consists of granite and other crystalline rocks of unfossiliferous sediments.

An average rainfall of about 100 cm received during the monsoon months and its erratic distribution further necessitate suitable watershed management interventions.

A calendar year is generally divided into three main seasons viz. winter (October-February), summer (March-June) and monsoon (July-September) with a brief spring (mid February-March) and autumn (late September-October). Winter temperatures generally remain below 5°C and precipitation in the form of both rainfall and snow result from the western depression. The cold wave sets the migration of the nomadic shepherds to warmer valleys in the Himalayan foothills.

Summer temperatures remain above 20°C especially during April-June. The relative humidity remains about 40 per cent and the occasional hailstorms are known to cause extensive crop damage especially to apple, plum, apricot and peach. Rainfall during this period amounts to approximately 30 per cent of total annual rainfall.

Monsoon results from the South-west monsoon and about 50 per cent of total annual precipitation is received during this period. The temperatures are known to drop by 2-4°C and the relative humidity reaches 60 per cent. In addition to precipitation the "directional aspect" is crucial for development of vegetation. The Northern aspect receiving less direct sun-shine and facing the snowline has consequently, lower temperatures and high moisture retention which inturn create ideal agro-climatic conditions for the cultivation of temperate crops, especially fruit crops. The southern and western aspects suffer high moisture loss since they receive more direct sunshine and consequently they support a poor vegetation cover. The difference in temperature on the different aspects of hills can also be explained in terms of differential isolation i.e. by proximity or distance from the equator, for example isolation factor for the southern aspect is about 1.5-2.4 times higher than that for the northern aspect.

Socio-economic Features

With horticulture occupying a prime position the land use system can be defined as horti-agri-pastoral. Forests account for about 25 per cent of the total geographic area of this zone. The forest cover varies from thick to sporadic. The chief species are kail, deodar, walnut and oaks etc. varying of course with altitude. The grasslands here are heavily grazed both by draught animals and by migratory graziers. The grass cover comprises mainly of *Themeda, Arunidnella, Hamertheria, Heteropogon* etc. and the legume component is mainly made up of white clover. While traditional millets such as Kongni,

Cheenee, Kodda and Bathu, etc. are fast disappearing. Cash crops such as potato and apple form the backbone of the economy with apple alone accounting for more than 78 per cent of the area under fruits. Pome and stone fruits are also grown in this zone. It is estimated that this zone alone contributes 96 per cent of the total temperate fruit production in the state. The traditional almost static farming has changed into a dynamic horticulture led system. Further, with the emergence of the market, farm strategies are no longer focussed only on security and precautionary motives.

Livestock composition is dominated by small ruminants due to availability of larger 'support' area. Cattle form the bulk of livestock rearing and are dependent on natural grass alone. Significantly, fodder is not grown. Hay, collected from natural grasslands is the main source of livestock feed.

The introduction of apple to this zone (in Himachal Pradesh) was the comer stone of the overall horticultural development strategy. Apple production accounts for nearly 40 per cent of the fruit area and about 90 per cent of fruit production. The economic transition was facilitated by the simultaneous development of supporting infrastructure by way of research and extension mechanisms, higher budgetary allocations, government policy initiatives (e.g. launching of the small farmers developmental agency; SFDA etc., an ever increasing focus on infrastructure e.g. roads, market development and Agricultural credit concentration, especially for capital demanding cash crops, in zone of the state is truly impressive.

For decades prior to economic transformation/transition societies in the both temperate and cold desert regions, as elsewhere, were part of economic conditions that centred chiefly around the security and precautionary motives. As such, production of hardy crops e.g. millets etc. was common practice. In the absence of structured markets as we have today, barter trade was resorted to. The absence of physical infrastructure, low economic resource base etc. translated people to adapt their lifestyles to suit their specific locational situations. Stated simply, subsistence farming with a focus on crops that met this goal was the common practice. These 'adaptations' ensured that agriculture was practiced on 'sustainable' lines with better income distribution and equity consequences. Undeniably, such farming/agricultural practices evolved under low population pressure in the different agro-ecological regions and depended entirely on locally available resource, yet, they do hold lessons for us even today.

This is to be appreciated in light of the realization that increasing populations and the consequent reduction of per capita resource availability dictate that a more efficient resource utilization regime be adopted. Such a regime is embodied in indigenous technological practices. Attempts to document such ITK also serve the purpose of preserving such knowledge for posterity, otherwise much of it would be lost due to the largely oral tradition of transferring such knowledge to successive generations. Such an exercise also has relevance for development agencies and extension services for providing relevant (technical) support for sustainable development and management of natural resources.

Soil and Water Management Techniques

In mountain watersheds, irrigation has been practiced as an art for about 3000 years now. Historical records bear testimony to the existence of a number of irrigation works in different parts of the country. In the Himalayas, the perennial river Ganges made it relatively easy to divert its flow through inundation channels. In the south, where rainfall is scanty, the practice of trapping rain water in large tanks and ponds for agricultural purposes is widely adopted.

From time immemorial, surface irrigation methods have been followed. The most effective irrigation method for a particular area depends on the slope of the land, the nature of the soil, the type of the crop and availability of funds.

In mountain areas, water continues to be the scarce commodity not only for irrigation but even for drinking and other domestic uses. This difficulty has been experienced very frequently, inspite of the fact that important rivers namely Sutlez, Beas, Ravi and their tributaries originate from these hills. The existing resources are further declining due to heavy biotic pressure and lack of management of existing resources. Most of our Agricultural/Horticultural activities are carried on under rainfed conditions and this require proper management of available water to be conserved for dry periods.

Sources of Irrigation Water

In the hill region, the scope of boring tubewells, canals and even lift irrigation is limited, such facilities are confined to the low laying areas. Therefore, the most common source of irrigation remains the small water channels locally called Kuhls which intact accounts for 85.83 per cent of the total area under irrigation in hills.

In cold deserts, some villages get water for irrigating their lands from some perennial torrents. In Spiti valley, the source of irrigation water is generally local nallas. Glacial water in cold deserts of Himachal Pradesh which forms the prime source of sustaining life in the region is brought to the field by making Kuhls (Water Channels).

In Kinnaur and other regions, the source of irrigation as well as drinking water is melting snow on the high peaks which runs downward in the shape of small and big nallahas (streams) and also spring out at certain points.

Construction of Kuhls (Water Channels)

In cold deserts of Himachal pradesh kuhls (water channels) are built along the hill gradient for maintaining proper gravity for irrigation. Kuhls are commonly found in West Himalayas cold deserts. The technique for the preparation of kuhls for irrigation purposes seems to have originated since Babylonian times, it is still one of the commonest ways of bringing water to the crops. If the river has a steep gradient, water is diverted into a canal some distance upstream and led along a contour so that it can flow to fields by gravity.

In dry temperate zone, kuhls (wooden water channels) are generally made by making notches at the natural water sources and the water is diverted to the fields for irrigation to different terraces, using the natural gravitational flow of water. Since the topography of the area consists of very high slopes and rocky terrain's, wooden water channels are used at many places as water passes from one place to another. The water channels are built and managed by the villagers with no government assistance. In the lower areas of H.P. bamboo pipes are commonly used as irrigation channels on depressions/small nala.

In west Himalayan cold deserts for the optimum harnessing of water for irrigation, water channels are constructed along the natural gradients. The irrigation channels (kuhls) are diverted from river tributaries by making use of the natural gradients thus the level of water is higher than that of the cultivated fields. In upper Kinnaur, the channels (kuhls) are simply dug in the ground to regulate the flow of water. However, where the digging of channels is difficult or the channel has to pass through a village path, underground channels covered with slates are constructed. However, in some parts the wooden channels are also used which are put like a bridge over the path. These channels are made by making a deep grove in the tree trunk or a thick branch.

Distribution of Kuhl Water in Fields

In the cold deserts of Himachal Pradesh, participatory management is employed for distribution of water. All disputes regarding the distribution of water through kuhls (water channels) are amicably settled without hampering the water requirement of any period.

In the West Himalayan cold deserts, all the irrigation channels (Kuhls) cannot be run satisfactorily due to non-availability of sufficient water from Nallas/Khads. This is because of scanty snowfall during the winter months. The majority of hamlets, which lie on the plateaus on the sides of main river get water from the streams which trickles down from the cliffs overhanging the plateaus. These hamlets are the worst off for water, for in the year of scanty snowfall, the streams dwindle quickly and dry up in the beginning of August. Additional snowfall in winter results in less water in natural springs during the season, whereas less snowfall in winter result in the reduction of level in natural springs during summer and consequently crop suffer.

Kuhls are a time tested community made water channels for sharing the glacial water for ensuring cent per cent irrigation in otherwise dry and porous soils.

In Spiti valley, the farmers have developed the irrigation water distribution system on the basis of their land holdings, in which every field is irrigated timely. So there is no dispute regarding the maintenance of kuhls and irrigation water distribution.

In Kinnaur and other regions, nallas passing through a village are harvested on turn basis called pala. Temporary channels are dug by the farmers towards their fields. The whole community is divided on the basis of number of farm families and one family gets one full water day to irrigate their fields turnwise. For example, if there are 20 farm families in a village, the turn falls after every 20 days. But two adjoining families may share the water for half day each when there is turn of either of the two families. This way these two families get a chance to irrigate their fields after a gap of 10 days rather than 20 days. This way the distribution of water is so well managed that maximum use of water takes place in a particular village. The turn of a family comes/starts around 2000 to 2200 hrs on a particular day and all the members of the family are engaged in the job on its turn.

In upper Kinnaur, the irrigation technique is much more pronounced. The fields are generally divided into small compartments by making earth bunds to allow water to stand in the field for a longer duration for saturating the soil. Hence need for second irrigation arises only after 20 to 25 days even in those agricultural crops which otherwise require irrigation after a gap of 10-15 days. At the first turn of irrigation, first compartment is irrigated; followed by second and so on. On the second turn of irrigation, however these compartments are irrigated in reverse order, i.e. sixth compartment is irrigated first followed by fifth and so on.

In temperate areas of cold deserts crop cultivation without irrigation is not possible because precipitation takes place in the form of snowfall. People take advantage of glacial water and perform collective operations for effective distribution and ensured supply of this scarce source. The management of water in a particular field is regulated by apportioning into different compartments because of the season. The mouth of first compartment is closed to regulate the flow of water towards the second compartment. The same method is adopted to irrigate the following compartment. This results in raising the height of channel in front of the first compartment than the channel in front of the second compartment and so on. Now when this field is irrigated during its second turn, the water flows straight towards the fourth compartment. This practice prevents the washing off the upper fertile layers during irrigation.

In the entire Spiti valley, the first irrigation is done 40 days after sowing of crop takes place during April. In the initial stage of watering from the Kuhl to field, the ladies bring water to the field by the use of Urma which is made from animals horn. As per the turn pertaining the Baraghar watering/irrigation is done by constructing small beds in the fields. This method is time consuming and laborious. But on the other hand this method checks the loss of nutrients by leaching. Uniform watering of the plants with equal flow, checks the nutrient loss from field to field and from one bed to another. In Ladakh and other regions standardized irrigation schedule for different crops is followed.

The general schedule is:

Irrigation number	*Stage of crop growth*	
	Local name	*English name*
I	Tol Chu	Germination
II	Sak Chu	Growing
III	Non Chu	Flowering
IV	Gep Chu	Seed setting
V	Do Chu	Crop ready for harvest

The “Gep Chu” or 4th irrigation depends on the colour status of the crop. If crop seems yellow in colour “Gep Chu” is delayed. However, the colour position is blackish “Gep Chu” is hastened.

The farmers have developed irrigation schedule matching the stages of crop growth. Thus, irrigation during critical stages results in the maximization of crop yield as well as water use efficiency.

Use of Kuhl Water for Running Water Mills

Kuhls are built along the hill gradient for maintaining proper gravity for irrigation and running water mills. Wooden water channels are also used for running water-flour mills. These wooden channels are generally made by making notches at the natural water sources and the water is diverted to the water mill, using the natural gravitational flow of water. Since the topography of the area consists of very high slopes and rocky terrain's, wooden water channels are used at many places as water passes from one place to another.

Granite stones are used for grinding food grains. Long wooden channel placed at steep gradient is used for maintaining the high speed of the water flow. This is necessary for maintaining the high speed of the water mills wheel.

Now a days water mills are very rare. Water mill technology is in an extinct stage, because of power supply availability and less grain production. Food is purchased from cooperative societies or private shops now a days.

Methods of Irrigation

Flooding of glacial water for higher crop productivity. In most Himalayan cold deserts water is brought in channels from glacial melts for irrigating the fields. Flooding the fields with the glacial water for improving crop productivity is also common.

The deposition of fresh silt with unweathered minerals (especially lime) forms glacier source of fresh salts. The glacier melted water is often below 2°C which protects the crop from different kinds of diseases.

Indigenous Drip Irrigation

The practice of using pitcher water as a source of irrigation on new fruit plantation in sandy loam/loamy sand soils, in areas of canty rainfall is prevalent in temperate districts of Himachal Pradesh. The pitcher is placed in soil and the new plant is planted close to it.

The pitcher is filled with water during summer months (April-June) and stone/slate lid is placed on the top. The roots draw moisture/water from pitcher which is turn reduces the mortality. The pitcher once filled, supply sufficient moisture for atleast two weeks and then again it is filled with water.

Bamboo Drip Irrigation System

In this system of irrigation bamboo channels (open) are used for irrigating the fields.This system is common in North-East regions of India. Small holes are made at the internodes of open bamboo channels, from where water gets trickled down in the field. These channels are placed along the natural gradients. In these channels, no uniform head for water trickling is maintained.

Manual Irrigation in Vegetables

In the initial stage of watering vegetables, people bring water to their fields with the help of buckets and in Spiti valley ladies bring water to their fields using Urma which is made from animals horn. In this method after bringing water in buckets, water is supplied to the vegetables with the help of lota (mug), whereas in case of Urma, irrigation is done by constructing small beds in the fields. But this method is too laborious and time consuming.

Water Harvesting Methods

Small Ponds for Spring Water Collection

Another method is the collection of spring water in small reservoirs scattered at intervals on the high uplands and then drawing water from these ponds when required. It is a common practice in cold deserts and temperate wet Himalayas. Water from these ponds is used for irrigating crops and also for drinking purposes.

Harvesting of Dew and Fog Water

In plains and in valleys occurrence of dew and "pale" is very common after the receding of monsoon. After monsoon the humidity remains quite high (85%) in the atmosphere. During night time, temperature falls down sharply resulting in the formation of more water molecules from vaporous. As they are heavier, they fall on soil surface and make the layer moist and wet.

In the hills, there is traditional practice to plough the fields early in the morning before dew or fog water is evaporated. By ploughing, moisture is mixed with soil particles in the plough layer i.e. 9"-12". This moisture is well retained by soil. If soil is clayey in nature, retention of water remains for a longer time and becomes a source of soil moisture. It is quite useful for land preparation in October-November and for the sowing of rabi crops like wheat, barley and pulses.

Roof Water Harvesting

In the lower areas of Himachal Pradesh during the rainy season, roof water is collected in dugout structures which are known as "diggi" in Kangra district and "'Khati" in Hamirpur and Bilaspur districts. These structures are dug in hard rocks. Not only roof water but also surface water is collected in dugout structures.

Harvesting of Rain Water

In the hills, rains are erratic and torrential. Relatively high percentage of rain water goes as run-off and stream flow. It carries fertile soil and plant nutrients which makes the soil degraded and barren. In some areas this excess water is stored directly in the farm ponds, depression or stream flow or is diverted to safer points where it is stored. The stored water in ponds and depressions is used for irrigational purposes, as a life saver or for supplementary irrigation during lean periods. It is also stored in dugout structures. In some areas during summer, it is used as drinking water humans, livestock and for other domestic purposes.

The ponds with time are sealed, with silt and clay particles thus infiltration/percolation losses are reduced and ponding time and volume of water is increased.

Harvesting of Water from Snow Melting

Harvesting of water is also done by constructing water ponds and water is collected in these ponds from melting snow.

Use of Pang (Spang) Grass for Controlling Seepage and Side Losses in Water Tanks and Irrigation Kuhls. In Ladakh Pang (Spang) grass is used as the inner lining of zings (water ponds) and irrigation kuhls for checking percolation losses. The use of spang grass which is growing profusely in Ladakh, explains its non-permeability properties similar to that of polythene sheet or cement lining. Its chemistry is required to be analysed, as the farmers claim its utility in water retention is far superior than the polythene/cement.

Moisture Conservation Through Mulching

In Kinnaur, covering the surface of soil with chilgoza tree needles and grass from the Kandas (hill tops) is a common mulching practice. Mulching conserves soil moisture in the fields. It also helps in the moderation of soil temperature. In this way hydro-thermal regime of soil is improved. However, the continuous use of chilgoza tree needles increases the acidity of the soil.

In the hilly areas, ploughing is done, which aids in moisture conservation, as the soil acts as mulch. In Ladakh, farmers regulate optimum irrigation by inserting a belcha (spade) in the soil If it is completely inserted (front portion), the land is considered to be properly irrigated. Similarly, in a few other cases, mud is thrown in the air. Its splitting into pieces shows proper irrigation. Complete insertion of the. front portion of belcha (spade) or throwing of mud in the air and its consequent splitting into pieces indicate the soil moisture level at field capacity, where 100 per cent moisture is available to the crops.

Drainage

During rainy season the rains are torrential, which causes splash erosion resulting in the sorting of particles and the formation of false compact layer on the surface. It yields water pounding and subsequently water logging. Crops such as maize, capsicum, tomato which are grown during this season are very sensitive to water logging. In our traditional agriculture there is a common practice that during the preparation of a field the slope of a field, is kept inside which is provided with a channel to take excess water from that field to a safer place, from where it is disposed to stream or nalla through grassed water ways. The grassed water ways are kept permanently and help in the drainage. These channels and grassed water ways are positioned in such a way that they do not hinder any agricultural activity such as ploughing, hoeing and harvesting.

Use of Smoke for Protecting Fruit Crops from Frost Damage

In the lower areas of Himachal Pradesh, Mango plants are mostly damaged by frost injury during winter months i.e. December and January. Smoke layer protects the mango plants from frost injury. This practice is common in the lower areas of Himachal Pradesh.

Soil Management

Cultural Practices

In West Himalayan region, in the month of March/April, when snow melts and weather condition improves, the bunds and comers of the fields are dug-out and weeds and grasses are removed with the help of spade and clods. The grasses or weeds are beaten up and then soil is separated from these clods and collected in lower fields.

This practice of removing weeds and grasses from bunds and corners by digging helps in weed control in the cultivated fields. Secondly area under crops remains the same as that of previous crop i.e. area is not wasted for weeds and grasses. Thirdly the soil added in lower fields from the bunds of upper field is rich in nutrients and it improves the soil fertility.

Use of Broader Plough in Upper Valleys

Ploughs are broadened in Ladakh by attaching flat wooden pieces to both sides of the iron blade. This indigenous plough is preferred over the one available in the market. This technology seems to have twofold functions of saving labour and that of stabilizing the loose sandy strata in one ploughing action, which suits the small terraces.

Sheet Erosion Control

It is not a damaging form of erosion, mainly because it is often not recognised and seldom treated. It accounts for the loss of billions of tonnes of soil every year. Due to splash of rain drops particles are knocked loose and then carried away by the runoff. The sheet erosion result into rill and gullies which are controlled by very cheap treatments. Sheet erosion is more apparent in forest areas that are devoid of ground cover or wastelands with very few standing trees.

There is traditional practice to keep surface maximum covered with grasses, shrubs etc., grazing is done in rotation and is allowed only during certain times. It is avoided during the flowering and seed setting stages of grasses. Fibrous rooted shrubs and grasses planted as hedges along the contour of the land slow the runoff, weaken the erosive power of water and cause it to deposit its load of valuable soil behind the hedgerows. As a result the runoff proceeds gently down the slope where hedges have been planted at the correct vertical interval without erosive effect In the foot hills, erosive capacity of stream flow is also reduced by spurs of loose boulders.

Traditional Rainfed Farming

In the hilly areas, most of the area is rainfed except for a few pockets in valleys where irrigation facilities are existing. The choice of crop and rotation, completely depends on crops which require less volume of water. For rotation, legumes are important as mixed crop. During rotation, when rainy season erosion permitting crops are grown, such as cowpea, 'kuth', these form integral part of the mixed cropping system. The crops are chosen as per their nutrition e.g. from old ages protein rich pulses are part of cropping pattern. The coarse grains like 'phaphra', 'chulai' are also grown very commonly which are very rich in nutrition.

Within the premises of the house it is mandatory to have fruit plants such as citrus, mango, anar which provide seasonal fruits rich in vitamin C., carbohydrates etc. The fields are well protected with biofence of thorny shrubs or their cut pieces. The traditional rainfed farming is done irrespective of land with respect to slope and other characteristics. There are chances of sheet erosion but with traditional knowledge, crop rotation is adopted in such a way that during peak runoff periods sowing of close growing crops provide protection to the soil.

Terracing

From old times, land in the hills has been put under cultivation on scientific lines as cultivation is done up to 25-100 degree slope, where there are many chances of landslips, sheet and gully erosion. But with 'bench terracing practices' the menace of soil erosion is controlled and is very common in hill fanning.

The terraces are constructed across the slope i.e. along the contour. The size of the terrace is decided by the prevailing degree of slope. The terraces are supported by risers of suitable heights and width. The height of riser is again decided by the degree of slope. The risers are sometime made of loose boulders supported by grasses. The roots of grasses help in binding and keeping the boulders intact at a place. The roots of grasses help in drainage of excess water. With the traditional knowledge, farmers are keeping the risers toward inner slopes. In paddy growing areas the risers are erected to facilitate the pounding of water in the field. This type of bunding and terracing is continuing from centuries and terraces are still intact. The bunds are again used for growing palatable grasses which is used as fodder for livestock and trees are meant for fuel, fodder and fibre. The examples are beul, shisham, mango etc. In lower areas the bench terraces are known as "khet". Sometime on the risers contour hedge of grass like khus, local grasses also established.

Use of Maddim (A Plain Wooden Structure) for Field Levelling

Maddim is used for levelling ploughed lands. A heavy stone is put on the maddim for increasing the pressure required for levelling. Sometimes, a man may also sit instead of a heavy stone.

Such an indigenous technology for field levelling is called planking. With this practice, there is very good seed soil contact and very good germination of the crops. Secondly, there is moisture conservation in the fields. Thirdly small soil clods are pressed and broken into finer particles and this way soil structure is improved.

Curved land ploughing for intensive land preparation; soil conservation and water retention. In west Himalayan cold deserts, ploughing is done in a curved (sword like) manner from the bottom to the top of the slopy land holdings.

Ploughing land holdings in a sword like pattern ensures proper land preparation which includes proper ploughing of the corners which otherwise would have remained unploughed. The ploughing of slopy lands from bottom to top also helps in soil conservation as it checks the loosened soil strata falling from the upper side to the lower. The curved pattern is useful in maintaining infiltration rate of water which otherwise gets wasted with sudden runoff.

Conserving Productive Soil Layer Against Wind Erosion

In west Himalayan cold deserts, fields are irrigated in autumn so that the top layer is prevented from being blown away. In spring the moistened soil eases ploughing. The productive soil layer, which is very thin, needs conservation against heavy wind erosion, a common feature of the cold deserts. This appropriate soil conservation technique also helps in easy and timely ploughing for meeting the requirement of short growing season. The moist upper layer of soil which gets frozen in winter also serves as a protection against wind erosion.

Cultivation of Levelled/Flat Lands for Preventing Soil Erosion

In west Himalayan cold deserts, cultivation practices are confined to the levelled/flat lands only.

This practice helps not only in the rational land use but also checks soil erosion in otherwise sandy and loose strata.

Contouring of slopy lands: Ethno-engineering for soil conservation

In west Himalayan cold deserts farmers have developed this technology for cultivation of slopy lands by constructing terraces comprising of plots and sub-plots by using small stones. Stone wall fencing is also constructed for individual land holdings. Terracing of slopy lands helps in conserving soil and moisture and prevents soil erosion. This also helps to carry out other field operations including proper use of irrigation water for checking the surface runoff.

Use of Loose Boulders Spurs for Reducing Soil Erosion

In some areas, people use loose boulders spurs for reducing *the* cutting effect of stream flow in a small nalla (Choes). Use of loose boulders diversion dam with spillway in centre for reducing soil erosion. This method is very common in lower areas of Himachal Pradesh. These dams are constructed across the streams for controlling soil loss.

Use of vegetation, live check of bamboo pieces and loose boulders.

In this method of vegetation, live check of bambo pieces and loose boulders are used for controlling gully erosion. This practice is also common in lower areas.

a) Vegetation

b) Live check

c) Loose boulders stabilized with grasses.

Use of River Bed Soils

The river bed soils are used for raising crops. These soils are rich in nutrients as nutrients are removed with soil from hilly slopes and are deposited in the river beds.

Soil Fertility Management

Proper soil management, ensuring continued maintenance and building up of fertility at a high level is indispensable for the profitable use of agricultural lands. While chemical fertilizers introduce extra concentrated supplies of readily available plant nutrients to the soil, the beneficial effect of organic manures predominantly lies in furnishing humus forming material to bring about improvement in the soil structure, water holding capacity, microbial population and its activity, base exchange capacity and resistance to soil erosion. Much of the plant food removed by the crops is restored to the soil through the application of organic manures.

Soil Management by Crop Residue Harvesting

This practice is prevalent in west Himalayan cold deserts. Barley and wheat stumps (in Zanskar) are pulled out by hand along with the complete root system. Soil is softened by a light irrigation a day before. Wheat is often pulled out while standing, but kneeling or squatting is practiced for barley. Handful of these plants are beaten up against the legs (occasionally a small apron is worn) to shake off most of the earth. These bundles are then piled up like the tiles of a roof. The ears of the lower row are covered and protected from birds by the roots of the upper stacks. In cold deserts of Himachal Pradesh, barley and buckwheat (in double cropping farming system) are also pulled out by roots. This helps in uprooting weeds, soil loosening and porosity maintenance for the coming crop. Other practice is to harvest crops as close to the grounds as possible. The roots are made to stay in soil for humus production. Very little plant material (stem and roots) is allowed to be left in the soil (in Ladakh) as a protective measure against the soil borne diseases. This practice also increases the fodder resource in winters. Retention" of roots in soil (in single cropping) contributes towards humus availability which improves the soil structure, porosity and water holding capacity of the soil.

Soil Mixing with Night Soils

This practice is prevalent in Ladakh and other parts. Soil with human excreta is mixed and broadcasted over the fields during winter months. Soil is collected from cultivated land holdings and particularly from field bunds of sub-plots for mixing.

The night soil/human excreta possess immense manurial potentiality as it contains the major plant nutrients like nitrogen, phosphorus and potassium. So the addition of night soil/human excreta along with soil from cultivated field improves the soil fertility. The practice of collecting soil from cultivated land and fields helps in easy ploughing during summer cropping.

Organic Manuring, Collection and Management

Organic manures derived from plant and animal resource, are valuable byproducts of farming and allied industries. Organic manures which is bulky in nature but supply the plant nutrients in small quantities are termed as bulky organic manures e.g. farm yard manure, rural and town compost, night soil, green manure etc., whereas those containing higher percentage of major plant nutrients like nitrogen, phosphorus and potash are known as concentrated organic manures e.g. oil cakes, goat manure, sheep and poultry manure, blood and meat-meals, etc. Flocks of sheep and goats, contribute towards tribal economy by way of milk, meat, wool and manure. These flocks when taken for grazing are tied with small bags which cover their anal parts so that the excreta falls right into the bag.

This region is highly sandy with low soil fertility status. The collection of dropping of sheep and goats by tieing bags is indicative of indigenous wisdom to meet out the shortage of manure. This manure of the droppings of sheep and goats contains 3% nitrogen, 1% phosphorus and 2% potassium. In Spiti valley, organic manuring is done once a year because of mono-cropping pattern in the months of September-October after the crop. The manure is broadcasted in the entire field, which is followed by ploughing for thorough mixing. The richest manure is called Chaksa which comprises of human excreta and is collected in separate dry latrine pit. The main reason for its nutritional value is that even the bones of animals are thrown in the excreta which adds phosphorus and calcium to the manure.

The daily per capita availability of night soil, human urine and nutrients contained in it is as under:

Particulars	*Faeces (g)*	*Urine(s)*
Quantity (natural condition)	133.00	1200.00
Quantity (dry)	30.30	64.00
Nitrogen	2.10	12.10
Phosphorus	1.64	1.80
Potassium	0.73	2.22

This data shows that night soil and human urine have a great manurial potential with regard to nitrogen, phosphorus and potassium. Due to this potential, it is considered good manure by the farmers.

Secondly cattle dung is collected in heaps within cattlesheds during winter months, so that it decomposes under relatively high temperature conditions. Then it is placed out in the open during summer in the form of heaps for further decomposition. Actually the cattle dung contains 0.2% nitrogen, 0.1% phosphorus and 0.15% potassium and cattle urine contains 0.6% nitrogen, 0.1% phosphorus and 0.5% potassium. Due to these immense manurial potentialities of cattle dung and urine, the use of this manure is very much popular among farmers.

Manuring is required for wheat, paddy and maize, which are the main crops of the Bharmour and Pangi regions. The traditional means of manure are as follows:

i) Dung of livestock, mostly cattle, collected from the sheds, pens and camps of livestock
ii) The leaves and grasses which were used as bedding for the animals, got soaked with the excreta/urine of livestock and were then collected periodically.
iii) Feeding of sheep and goats in the fields: This method of manuring is very much in vogue in those places which are visited by the Gaddi graziers, whether enroute to their camps or on move with their herds. The Gaddis are paid for this benefit. These traditional practices continues unchanged. The only improvement that has been made is that the heaps of cow dung are well covered with something or the other in order to protect them from rains and snow.
iv) In the wet temperate Himalayas, green and dried pine needles are collected in heaps and used as bedding material. Before using as bedding material these pine needles are cut into small pieces.

In the absence of chemical fertilizers, organic manuring is the chief mode of soil fertilization. All efforts are made to collect and use animal dropping and for their subsequent decomposition along with the leaves and grasses which are used in manuring the crops. This is the traditional organic manure and is most readily available to the farmers.

It is the product of decomposition of the liquid and solid excreta of livestock, stored in the sheds, pens and camps of livestock along with varying amounts of straws or other litter used as bedding. This farm yard manure/compost prepared from farm litter, liquid and solid excreta of livestock contains 0.5% nitrogen, 0.2% phosphorus and 0.5%

potassium. To enhance the productivity, people in Kinnaur still use the farm yard manure. It is worth mentioning that here animals are kept primarily to meet the need of manure.

Donkeys, cows, goats and sheep are the main source of manure. The manure is collected either from the cowsheds inside the house or the cowsheds outside the house. Generally, the ground floor in each house is used as a cowshed so that animals can be looked after in a better way during winter months. The dung is put outside the house in a heap form in lower areas, whereas, in upper areas, it is directly put in small heaps in the fields. These small heaps of dung are covered with a thin layer of soil to avoid the dispersion of manure by wind. The manure is directly mixed with the soil while ploughing. Farm yard manure is transported to the fields in Kilta (bamboo container) by people's participation and also by horses. Amongst the manures, the cowdung is preferred the most. According to most farmers the sheep and goat dung may lead to burning of crops if applied in excess. Ass dung though used is not preferred much. On an average 125 to 250 qtls of manure is used per acre by the farmers throughout the Kinnaur region. The practice of keeping small heaps of manure in open field with soil coverage in high altitude zones helps in better decomposition due to the maintenance of better temperature conditions. Use of sheep and goat manure in large quantities leads to burning of crops. The burning of crops is due to the toxic effects of high levels of nitrogen, phosphorus and potassium in goat and sheep manures. The goat and sheep manure contains 3 % nitrogen, 1 % phosphorus and 2 % potassium.

Use of Ash in Ladakh

Nutrient Recycling

The inhabitants of this entire region use cattle dung, shrubs and bushes as the main source of fuel. Ashes available, there upon, are mixed either with household waste or human excreta. Sometimes ashes are also broadcasted in the fields. Mixing of ash with household waste and human excreta aids in nutrient availability and recycling. Ash primarily meets the deficiency of potash. Availability of phosphorus is also ensured. In addition to this, human excreta and household waste also contains good amounts of nitrogen, phosphorus and potassium.

Softening of Hard Soils

In Nubra valley, hard soils are softened by putting ash obtained from cowdung, sheep/goat manure, fuelwood etc.

Through this practice upper layers of soils are not only softened but their fertility status is also improved, as ash contains phosphorus.

Increased size of potatoes through the use of ash and goat manure. A mixture of kitchen ash and goat manure is used in kitchen gardens (Nubra valley) for growing potatoes. The spreading of this mixture as an organic manure, increases the size of potatoes on account of optimum supply of nutrients in otherwise nutrient deficient soils. Secondly organic manure improves the soil structure, porosity and water holding capacity of the soils. In this way there is an overall improvement in physical, chemical and biological properties like microbial population etc., which has increased the size of potatoes.

Poultry Manure and Ash for Increased Vegetable Production

This specific technology is used only in case of tomato, brinjal, capsicum and cauliflower. Kitchen ash and poultry manure mix enhances vegetable production levels.

Stage of Farm Yard Manure in Cultivated Fields

In west Himalayan cold deserts, FYM with a thin coverage of soil is kept in small heaps in the fields from October to March. With the onset of summer months it is spread in the open field. Coverage of organic manure (FYM) with soil in open fields throughout winter helps in regulating (heap) temperature necessary for proper decomposition of FYM.

Green Manuring

In Bharmour area the practice of green manuring is localized in a few villages (paddy growing). Leaves and twigs of wild bushes such as basuti and kaimal are used.

Use of Goat Manure

In Ladakh, goat manure is considered to be more nutritious. Goat manure when added to millet fields improves production. Goats are specially penned in these plots/fields. Goat manure improves not only the millet production but also its taste. According to farmers vegetables grown in goat manure have longer keeping quality. It is easy to plough fields manured with goat excreta. Actually with the addition of goat excreta, there is improvement in the physical properties like soil structure, water holding capacity and porosity. There is also an improvement in soil fertility as it contains 3% nitrogen, 1% phosphorus and 2% potassium.

Use of Sachik Soil for Higher Crop Yield

Yellow soil (Sachik) found in Tagloom area is used as manure for enhancing crop production. Yak loads of this yellowish/dark brown coloured soil are scattered in the fields.

Biofencing with Seabuckthorn (Hippophae Rhamnoides)

This practice is prevalent in Spiti and other regions. There is a common practice to provide biofencing with seabuckthorn in cold deserts in general and Spiti in particular. The biofence of seabuckthorn being thorny in nature protects crop from stray animals. Its multipurpose utility as a nitrogen fixer, checks against soil erosion, conservation of soil and moisture, source of fuelwood and indigenous drug (rich source of vitamin C) makes it a promising plants for eco-economic rehabilitation of the region.

Sprawling of Ash Dust in Cucurbits and other Vegetable Crops

In the west Himalayan cold deserts, ash dust is a product obtained after the combustion of fuelwood. It has been observed that dusting of material in the fields enhance early maturity and high yield of vegetable crops. The reason for the early maturity of cucurbits and vegetable crops is due to the fact that ash dust contains sufficient quantity of phosphorus in available form to the plants. Secondly, in cucurbits the ash dust has been used to repel the insect pest of the crops. Thirdly, amendments of ash dust in the soil, improves soil structure and fertility. Ash dust is also useful in enhancing the maturity of bulb crops which normally takes 6-7 months for obtaining economic yield.

Drought Power According to Soil Texture

In west Himalayan cold deserts, ploughing is generally carried out by dzos, however in sandy situations horses are employed for its speedy completion. In Turpuk of Nubra valley ploughing is done by a single horse. Sandy soil have less soil strength than clayey soil. Due to this reason, the drought power requirement for ploughing varies according to soil texture.

Chapter 3

Surface and Groundwater Resources of Arid Zone

Water is a precious resource that is currently in need of long-term planning for storage and judicious use for the survival of mankind. Excessive exploitation of water without any systematic conservation programme can lead to depletion of the resource beyond recovery, as is being currently experienced in the arid regions. The water resources of the two most problematic arid areas in the country, the hot arid areas of western Rajasthan and the cold arid areas of Ladakh district in Jammu and Kashmir.

Water Resources of Arid Western Rajasthan

Western part of Rajasthan state is the most problematic area within the hot arid zone. The average annual rainfall here is 318 mm, as compared to 531 mm in Rajasthan state. Due to low and erratic rainfall, replenishment of exploited water is also very poor. During the Twentieth Century, different parts of western Rajasthan experienced agricultural drought once in three years to every alternate year. The overall probability of drought for the state is 47%. The weather condition even in average years for most part of the year remains too dry and inhospitable for successful growth of crops. Under such conditions of uncertainty, conventional cropping is risky and has necessitated widespread use of groundwater for irrigated cropping. Although the whole of Rajasthan state is categorized as water-scarce (having per capita water availability below 1000 m^3 year-1 Narain *et al.*, 2006a), the condition in western Rajasthan is more precarious. The west-central part of western Rajasthan is devoid of any drainage network and has meager surface water resources, which adds to the problem. Rapid urbanization and industrialization, increasing pollution of the water sources by the industrial units and over-extraction of water from deep wells has led to water quality problems. Despite climatic adversity, this is one of the most densely populated deserts of the world.

Long-term statistical analysis of rainfall data of the region indicates an asymmetric average storm intensity profile for storms of short duration, with the highest intensities falling in the first part of the storm. Occurrence of water flow in the watercourses is unpredictable, of short duration and high variability. The instantaneous discharge-duration curve shows very high and irregular peaks indicating the problems of controlling runoff. Regulation of sporadic discharge by means of surface reservoir presents many problems, especially the high ratio of storage to mean annual runoff volume required to produce the degree of control necessary for economical agricultural development (Goyal and Vittal, 2008).

Hydrological Zones

Hydrologically western Rajasthan has been divided into three broad zones (Venkateswarlu *et al.*, 1990).

Zone – I : Region with major input of surface water from more humid region, frequently with extensive irrigated agriculture. About 60% area of Ganganagar district in the north and 50% area of Bikaner district and 25% area of Jaisalmer district in the northwest lie in this zone. This is the main canal irrigated zone in arid Rajasthan.

Zone-II: Plain lands with a primitive or no stream network. The region has a system of repetitive micro-hydrology. Churu, Jhunjhunu, Sikar, Nagaur, Jodhpur and parts of Bikaner, Jaisalmer and Barmer districts come under this category. This zone occupies 52% area of arid Rajasthan.

Zone-III: Sloping region with an integrated stream network. The Luni basin, occupying the districts of Pali, Jalore, part of Jodhpur and Barmer districts, lie in this zone.

Surface Water Resources

The inherent surface water resources of the western Rajasthan are scarce and because of low and erratic rainfall, replenishment of these water resources is also very poor. Due to high atmospheric temperature and low humidity, a large part of the rainwater is lost as evapotranspiration. Except in canal command area in north, surface water potential is very low in the central, western and southern parts. More than 50% area of northwest Indian arid zone comprises of sandy plains, dune systems, eroded rocky/gravely surfaces and isolated hillocks. In central and western parts, the run-off generated in response to some high-magnitude rainstorms gets lost in sandy terrain. As per initial estimate approximately 280 x 106 m3 surface water is available annually for utilization in this part (Jain, 1968). Mehta and Kashyap (1970) estimated the surface water potential of this region as 200 x 106 m3, out of which 130 x 106 m3 is utilisable resource.

Sharma and Vangani (1992) estimated the surface water potential of this region as 1360 x 106 m3 out of which 47% was utilized till 1988. Nearly 33% area of the zone is occupied by the Luni Basin, the Sahibi Basin and a few smaller river basins. These are ephemeral drainage systems and convey runoff only in response to torrential rainfall during the monsoon season. The estimates of surface water resources for Luni basin ranges from 518 x 106 m3 (Dhir and Krishnamurthy, 1952), to 571 x 106 m3 (Dhruvanarayana *et al.*, 1964), to 868 x106 m3 (Mehta, 1970), to 939 x 106 m3 (Mehta and Kashyap, 1970), to 858 x 106 m3 (Anon., 1980), to 1130 x 106 m3 (Sharma, 1991). Large numbers of tanks, reservoirs, minor irrigation dams and check dams have been constructed at different locations in Luni basin and other areas to store runoff water during monsoon. About 550 storage tanks in the capacity ranging from less than 1.51 to 208 x 106 m3 are functional with total utilizable capacity of nearly 1169.28 x 106 m3 for providing irrigation to 0.102 x 106 ha land (Khan, 1997). Out of these, six reservoirs viz. Jaswantsagar, Sardar Samand, Jawai, Hemawas, Ora and Bankali, are the major irrigation tanks with capacity of irrigation of more than 4000 ha each. Jawai is the main source of drinking water supply to many towns and villages (Source: Irrigation Department, Govt. of Rajasthan).

About 15% area in the north-west arid zone of India receives major input of water through an extensive canal network. This region comprises of the arid districts within the states of Haryana and Punjab and the northwestern part of Rajasthan. The canal systems in this region are the Gang Canal, the Bhakra Canal and the Indira Gandhi Canal bringing water by diverting the flows of the Sutlej, the Beas and the Ravi in Punjab. The estimated designed capacity of these canals varies from 314 to 524 cumec (Murthy and Gulati, 1978; Uppal, 1978; Gupta, 1987; Kapoor and Rajvanshi, 1977). On the basis of average flow from 1982 to 1996 the canal water available in the four districts of Ganganagar, Hanumangarh, Bikaner and Jaisalmer is at the rate of 0.90 m ha-1 (Anon., 1988; 1996), far more than the stipulated mean value of 0.51 m ha-1. Indira Gandhi Nahar Pariyojna (IGNP) The Indira Gandhi Nahar Pariyojna (IGNP) receives its supplies from the Indian share of waters of Indus river basin as per Indus Water Treaty of September 1960 between India and Pakistan.

The three eastern rivers of the basin (namely Ravi, Beas and Sutlej) have been harnessed since then through several facilities dams, reservoirs, links and canals. Although water resources development of the Indus waters dates back to centuries, major activities started in the Tenth Century and continued to meet with the growing needs of the agricultural community. The Gang canal system built during 1922-1929 utilizes 1.11 million acre feet (MAF) of water which constitute the pre-partition share of India, drawn from the Sutlej river near Ferozpur Head Works.

RD 45 of the Gang canal was linked with Harike Barrage constructed in the upstream at a later date. The inter-state agreement for sharing waters of the Ravi-Beas rivers came into being in the year 1955 between erstwhile State of Punjab, Jammu and Kashmir (J&K) and Rajasthan, followed by Bhakra Nangal agreement in 1950. Agreements were later revised and finally as per 1981 agreement total share of water for Rajasthan in Ravi-Beas waters was 9.71 MAF. This quantum of water was further allocated by the Government of Rajasthan to IGNP stage I and II, 3.59 MAF(inclusive of 0.22 MAF for drinking and other uses) and 4.00 MAF (inclusive of 0.65 MAF for drinking and other uses), respectively. The remaining water was given to Gang Canal 1.44 MAF (1.11+0.33), additional water for Bhakra system from Ravi-Beas waters 0.21 MAF and Sidhmukh/Nohar Project 0.47 MAF.

Stage I consists of a 204 km long feeder canal, having a headworks discharge capacity of 460 m3 sec-1, which starts from Harike Barrage. 170 km of the feeder canal lie in Punjab and Haryana and 34 km in Rajasthan. The Stage-I also consists of 189 km long main canal and 3454 km long distribution system, which are concrete lined, and serve 553 kha of culturable command area, out of which 46 kha are served by pumping to a 60 m lift, through four pumping stations. IGNP Stage II comprises onstruction of a 256 km long main canal and 5,606 km of a lined distribution system, and will serve 1,410 kha of CCA (873577 ha area in flow and 537018 ha under lift), utilising 4,930 Mm3 yr-1 of water. The main canal in the entire length was completed in the year 1986. The irrigation intensity initially planned for Stage I and Stage II was 110% and 80%, respectively (GOR, 1999). The canal system was initially designed at a water allowance of 5.23, 3.0 and 2.0 cusecs/1000 acres for Stage I, Stage II flow area and Stage II lift areas, respectively (which works out to 0.36, 0.21 and 0.14 cumecs per 1000 hectares).

Irrigation in canal command area needs proper management to avoid water logging, soil salinity, etc. More than 60% of canal command area has sandy soils with poor water holding capacity. Nearly 50% of additional water applied in form of irrigation goes as deep percolation and joins groundwater. The problem of water logging is more apparent in recent years. The mean rate of water table rise varies from 1.1 m year-1 in Stage-I to 0.81 to 0.85 m year-1 in Ghaggar plain and Bhakra command, respectively, and 0.64 m year- 1 in Gang canal command area. On the basis of annual rise in water table, it has been found that an area of 1456 km2 has already turned critical (water table within 6 m of land surface). However, a far more serious water logging problem awaits Stage-II, owing to an underground hard substratum of gypsum within 10 m depth. In about 34% area (1205 km2) of the gross command area of 3544 km2, water collected in low lying areas does not seep down (Rahmani and Soni, 1997). Due to capillary action, the water comes to the surface with dissolved salts and evaporates, leaving the salt behind, thus making the land saline. According to one study (Chouhan, 1988), if surface drainage is not introduced in the waterlogged area, thousands of hectares of land will be submerged and salinised in 25 to 30 years.

In the adjoining Haryana state, Sirsa and Hisar districts have 148 km2 (3.0%) and 266 km2 (4.0%) area, respectively, within the influence of high water table (Rao *et al.*, 1986); the rate of water table rise is in the range of 0.14 to 1.0 m year-1. In addition to salts contributed by groundwater nearly 2.0 x 106 t of salts are added annually through canal irrigation. In Punjab state nearly 1.05 x 106 ha area is irrigated through canals in the arid districts of Faridkot, Firozpur and Bhatinda (Sidhu *et al.*, 1991). As a result about 80% of the total salt affected and waterlogged area of the state occurs in these districts; the water table is rising @ 0.52 to 0.75 m year-1.

Groundwater Resources of Arid Rajasthan

Groundwater resources of this region are very poor in terms of quantity and quality. Groundwater in this region is not sufficient even for drinking purposes. Over and above insufficient quantity, the groundwater is moderately to highly saline over large area. A dominantly sandy terrain and disorganized drainage network (drainage density is as low as 0.3 km km-2), and recurring droughts constantly exert pressure on already meager groundwater resources. The stage of groundwater development has exceeded 100% in Barmer, Jalore, Jhunjhunu, Jodhpur, Nagaur, and Sikar districts. Number of Safe blocks has been significantly reduced because of meager rainfall and over exploitation of groundwater resources mainly for irrigation.

Groundwater Depletion

Due to over mining groundwater levels are declining in 9 out of 12 districts of arid Rajasthan since 1984. Groundwater table in Jalore and Pali districts shows a decline rate of more than 0.50 m year-1. In Jodhpur, Jhunjhunu, Nagaur and Sikar districts groundwater decline rate is 0.44-0.48 m year-1. In Barmer, Churu and Jaisalmer districts the rate of decline is less than 0.20 m year-1. In Bikaner, Ganganagar and Hanumangarh districts water level shows a rising trend due to IGNP.

Future Groundwater Scenario

Future projection of groundwater utilization has been worked out for the year 2010, 2015, 2020 and 2025 considering the average growth rate @ 3.20% compounded annually, though the present actual growth rate is higher.

The projection presents a grim situation even when the numbers are on a lower side. The stage of development is likely to reach 161.6, 189.1, 221.4 and 259.1% in the year 2010, 2015, 2020 and 2025, respectively. Arid Rajasthan has to face the adverse effects of overexploitation. Strategies for Water Resource Management in Arid Zone Rain is the principal source of water in this region, which augments soil moisture, groundwater and surface flows. Agriculture and several other economic activities in arid areas depend on rain. Of the total water use about 85% of water is used for irrigation and remaining 15% is used for drinking, industrial and other purposes. About 65% of irrigation water and 30-40% of drinking water is subjected to serious losses. Hence, increasing water use efficiency, coupled with increasing availability of water through rainwater harvesting and management, is key to survival on sustainable basis. Rainwater harvesting, its conservation and efficient utilization can solve problem of water scarcity to a greater extent. Rainwater harvesting in small ponds (*nadis*), underground tanks (*tankas*), *Khadins* (Low lying areas) etc., is an age-old tradition here. These traditional structures vary in design, shape and size. These structures are now partially or sometimes totally neglected because of increased dependence on tubewell, tankers, canal water supply, etc. The traditional methods require revival and improvement for being more economical and efficient.

Rainwater Harvesting in Nadis and Ponds

The people of rural arid areas live in scattered settlements called *dhani's* distributed over sand dunes, interdunal plains and undulating landforms. Under such conditions it is inconceivable that organized water supply will be feasible to fully meet the demand of thirsty land, human and livestock. Rainwater harvesting in *Nadi* and farm pond for groups of farm families or community are the most viable proposition. *Nadi* is a dugout pond used for storing runoff water from adjoining natural catchment during rainy season. High evaporation and seepage losses through porous sides and bottom, heavy sedimentation due to biotic interference in the catchment and contamination are its major bottlenecks. Complete control of seepage and evaporation losses is very costly and not foolproof. Nadi can play a significant role in preventing complete crop failure (Mann and Singh, 1977; Singh, 1983). Singh (1986) reported better utilization of nadi water for raising nurseries, orchards and to support the initial establishment of trees observed instead of watering field crops.

Chatterji *et al.* (1985) reported pollution of nadi water due to free access of human and livestock, leading to the growth of many harmful bacteria and other water-borne diseases and therefore, not safe for human consumption in Nagaur district. Sedimentation in pond is another major problem. Shankarnarayan and Singh (1979) observed reduction in water surface area and drainage basin area up to 1.8 to 2.4 and 6 to 8 Times, respectively, due to biotic interference.

To overcome these problems CAZRI has developed designed *Nadis* with LDPE lining on sides and bottom keeping surface to volume ratio 0.28 and provision of silt trap at inlet (Khan, 1989). *Nadis* also help in recharging groundwater aquifers although their effect varies depending on the underlying soils and rocks. Where the substrate is rocky, it is estimated that they contribute a depth of 0.06 metres of water a year c ompared to 1.58 metres in sandy plains. A study of a 2.25-hectare *nadi* with a storage capacity of 15,000 cubic metres in the north Gujarat alluvial area calculated that the pond contributed as much as 10,000 cubic metres of water to the groundwater aquifer in one rainy season.

Farm Pond is an improved version of *nadi* with treated catchment and surplusing arrangement for removal of excess water. A farm pond of 20,000 m3 capacity was constructed at Kukma watershed at Bhuj in Gujarat by CAZRI in year 2004. Construction of farm pond resulted in assured availability of 20,000 m3water even in during 150 mm rainfall (Narain *et al.,* 2006 b). The collected water was used to provide irrigation to Datepalm, ber, aonla and other fruits plants in nearby area. Construction of large number of rainwater harvesting nadis and farm ponds can solve the problem of uncertainty of occurrence of rainfall and can store water during heavy rainfall for non-monsoon period for human, livestock and crops on sustainable basis. Therefore, construction/

renovation and desilting of nadis/farm ponds during drought relief measures by state government and NGO's can be beneficial.

Rainwater Harvesting in Khadin for Crop Production

Recurring droughts and long dry spells are regular feature of arid zone of Rajasthan, which result in crop failure or severe reduction in crop growth and yield. A traditional practice of *khadin* farming in arid Jaisalmer district ensures better moisture conservation and cropping. The system is very effective even where annual average rainfall is less than 200 mm. Khadin farms were first constructed by Paliwal Brahmin of Jaisalmer district in the Fifteenth Century (Sehgal, 1973). The Paliwals connected most of the local catchments into well-knit system of *khadin* farms for assured crop production even under low rainfall. Kolarkar *et al.* (1983) observed 2 to 16 times lower electrical conductivity (EC) of Khadin soils compared to outside farms. The reduction in EC has been attributed to leaching of salts through seepage water in khadin. The *khadin* soils hold soil moisture to last up to growing season. The average yield of 20 to 30 q ha-1 for wheat and 13 to 25 q ha-1 for chick pea without any specific agronomical practices and fertilizers were also reported under khadin. Tiwari (1988) has linked the *khadin* cultivation with the farming system of Kalibanga and Rang Mahal cultures of the Indian desert, which are presumed to be of Harappan age (3000 BC).

CAZRI has evolved a design package and guidelines for construction of khadins (Khan, 1992 a). Improved khadin has been constructed by CAZRI near village Danta in Barmer district. The catchment area of the khadin is 137 ha with 6.88 ha submergence. Provision of 40 m bed bar in 450 m long earthen embankment was provided for spilling over excess water in khadin bed. The total water storage capacity of khadin is 54.2 x 104 m3 and beneficiaries are four farm families (Khan, 1998 a). A Khadin of 20 ha area was developed in Baorali-Bambore watershed with surplussing arrangement. Before construction of Khadin, uncontrolled runoff from upper catchment used to wash away seeds, fertilizers, and standing crops, besides loss of valuable water. After construction of Khadin, farmers could take excellent Kharif and Rabi crops (Narain and Goyal, 2005). Collecting water in a khadin aids the continuous recharge of groundwater aquifers. Studies of groundwater recharge through khadins in different morphological settings suggest that 11 to 48% of the stored water contributed to groundwater in a single season. This replenishment of aquifers means that subsurface water can be extracted through bore wells dug downstream of the khadin. The average water-level rise in wells bored into sandstone and deep alluvium was 0.8 metres and 1.1 metres, respectively. There were 500 such *khadins* covering an area of 12,140 ha in Jaisalmer and the crop production from such areas was adequate to feed the people of Jaisalmer district. Largescale development of *khadin* farms at suitable locations in western Rajasthan can enhance land productivity.

Rainwater Harvesting from Constructed Catchment

Under this system, catchments are constructed in such a way that runoff is directed to a desired destination instead of spreading here and there. The desired destination may be crop/tree to supplement the soil moisture or a reservoir for storage and subsequent utilization. Inter-row, inter-plot and micro-catchment are some of the constructed catchments, which can be used to enhance the availability of water. Runoff from microcatchments was generally found to depend upon rainfall and catchment characteristics.

The results of seven years' field studies showed that microcatchments produced 13 to 32% of rainfall as runoff at 0.5% slope; and even higher amounts at 5% (36 to 45%) and 10% (26 to 44%) slopes. Runoff generally increased with decreasing slope length; runoff for 5 and 10% slope were nearly equal but greater than for 0.5% slope (Sharma, 1986). Apparently, there is a critical slope beyond which runoff is not affected by slope increases. A large part of the rainfall is absorbed by the sandy soil of the catchment, thereby reducing the total amount of harvested water. Various surface covers and sealants were experimented to increase runoff. Plastic covered catchments were found to generate 95% runoff while Janta emulsion (asphalt), pond sediment and compacted catchments yielded 91%, 88% and 66% runoff, respectively (Sharma *et al.*, 1986). The results of studies at farmer's field in Kalyanpur (Barmer district) showed very high efficiency of moisture conservation with stone and sand filled polythene bags with associated improvement in growth and establishment of Ziziphus mauritiana (ber) seedlings (Ojasvi *et al.*, 1999). Inter-row water harvesting using ridge-furrow has also been found useful in raising dryland crops.

Under this technique, 50-60 cm wide ridges alternated with 30-40 cm inside furrows (15 cm deep) are constructed using ridger equipment. Crops are planted in furrows, adopting a paired row design. Ridges yield runoff to the furrows, thus enhancing the moisture regime in the root zone. Singh *et al.* (1973) reported 210% increase in the yield of pearl millet with this system. They concluded that ridge-furrow technique has better adaptability for small holders, as no area of the field is lost to catchment construction. Singh (1982) suggested maintaining the furrows and ridges as permanent structures.

Thus, if the tillage is restricted to furrows only, the energy input can be substantially reduced. Rainwater harvesting for groundwater recharge As per groundwater estimates of Rajasthan for year 2001, total annual groundwater availability is 11,159 mcm as against total annual water demand (draft) of 11,626 mcm. The overall stage of groundwater development for Rajasthan state is 104%, which is categorized as 'overexploited'. Presently out of total 32 districts, 14 districts are in the category of overexploited, 4 in critical zone, 8 in semi-critical zone and remaining 6 are considered in safe category. With increasing demand for water, more and more blocks are likely to be in the category of overexploited and need immediate attention for recharge of groundwater. Percolation tanks, pondage in stock tanks with infiltration galleries, sand filled dam, anicuts across stream, sub-surface barriers, etc., are used for groundwater recharge (Ojasvi *et al.*, 1996; Goyal, 1999). Sub-surface barriers constructed across ephemeral streams trap sub-surface flow to recharge groundwater aquifer (Khan, 1998 b). Construction of two subsurface barriers of 10 m length each within 300 m from water supply well was found enough to store runoff water required for a village having a population of 500 (Anon., 1974). Singh *et al.* (1989) reported that soil conservation practices have increased recharge to the extent of 14.02 to 19.52% of rainfall in Udaipur region. Adoption of conservation measures like anicuts, loose stone check dams, brushwood check dam, etc., in Jhanwar watershed (Jodhpur district) resulted in recharge/increase of groundwater level @ 0.33 m-.75 m year-1 (Bhati *et al.*, 1997; Goyal *et al.*, 2007). In another watershed at Osian-Bigmi (1991-96), conservation measures like loose stone check dams, vegetative barriers and anicuts resulted in rise in water table by 1.1 m, indicating the effectiveness of conservation measures for recharge of groundwater (Gupta *et al.*, 2002).

Rainwater Harvesting through Tanka for Potable Water

Good quality potable water is a global issue, particularly in developing world because 80% of the diseases in the world is due to poor quality drinking water. The problem of poor quality groundwater used for drinking is more acute in Rajasthan. Concentration of fluoride ranges from 0.4 to 90 mg l-1 leading to various diseases like dental fluorosis, skeletal fluorosis and, non-skeletal manifestation etc.

Rainwater is the purest form of water. Appropriate harvesting of rainwater from roof top and open and its utilization can alleviate problem of fluoride to great extent. Studies conducted at CAZRI have revealed that roof made of different materials can generate 50- 80% runoff that can be stored in underground cistern (*tanka*) which could provide excellent drinking water round the year. Rainwater can also be harvested in tanka using artificially prepared catchment. Collection and storage of excess rainwater in tankas is an age-old tradition for meeting domestic water requirements. Till today most villages depend on these structures as source of drinking water. Traditional *tankas,* constructed with lime plaster, typically have a life span of three to four years. They suffer from seepage and evaporation losses and in the absence of proper silt traps and pollutant-free inlets, the quality of the conserved water deteriorates over time, making it unsafe for drinking. In many situations, degradation of the catchment area means that it does not yield the quantity of water required to continuously replenish the structure.

Increasing Water Productivity by Reduction in Water Losses

Evaporation accounts for over two third of water losses from surface water bodies in hot arid regions. Reducing surface area by increasing storage depth can appreciably reduce water losses. The surface area can also be reduced by storing the water in a compartmented reservoir, and pumping the water from one compartment to another as the water is used, so that there are some full compartments and some empty, instead of a single shallow sheet when the reservoir is partly used. Covering the field with any kind of mulch also helps in reduction of evaporation losses from the surface. Evaporation loss can also be minimized by reducing wind velocity through shelter-belts of suitable tree species around water bodies or by artificially shading of water surfaces. Studies on artificial shading of water surface have shown encouraging results in controlling evaporation losses. Shading of water surface with polyethylene sheet successfully reduced evaporation by 91%. Evaporation reduction with floating materials ranged from 37% for *Saccharum munja* to 82% for polystyrene sheet. Foamed rubber sheet, polyethylene sheet and bamboo reduced evaporation by 74%, 66% and 53%, respectively. The floating polystyrene sheet and polyethylene covers were the most economical (Khan *et al.*, 1990). The micro-irrigation systems like drip and sprinkler economize both water and fertilizers. These systems could be popularized to increase the productivity of limited rainwater.

Flash Flood Management

Although most parts of arid Rajasthan are drought-prone, flash floods are not uncommon in this region. Since rainfall in this region is of convective nature and usually occurs at a very high intensity for shorter duration creating flash flood, particularly in urban areas where buildings and roads generate very high runoff even for little rainfall. During 1979 large parts of state witnessed flash flood due to very heavy downpour of more than 500 mm in just 3-5 days which cut off the state from rest of the country for several days. Between 1901 and 2003 western Rajasthan witnessed nine moderate and 19 severe floods. Although floods are considered a natural calamity, however, if the excess flood water or its part can be managed and utilized for rejuvenating the depleted aquifers as well as improving surface water resources the problem of water scarcity during droughts can be solved.

Harvesting and conservation of floodwater to rejuvenate the depleted aquifers by adopting artificial recharge techniques will remarkably improve the water availability for growing population. Out of 12 district of arid Rajasthan, Bikaner, Hanumangarh and Ganganagar districts have reported a rise of groundwater table due to canal irrigation and remaining 9 districts have recharge potential of 30.3x109 cum based on groundwater depletion taken between 1984-2003. Narain *et al.* (2006 a) have estimated a potential of 5.9 x 109 cum of flash flood for western Rajasthan. A part of flash flood can be used to recharge groundwater of these districts. Bilara limestone, Lathi Sandstone, Jodhpur Sandstone and Alluvium aquifers are most suitable for groundwater recharge.

Watershed Management for Soil and Water Conservation

Initial approach of conservation programs was based on particular land problem and individual holdings. In this approach often the upper reaches of the fields were left untreated and runoff from uplands used to damage the conservation work laid out in the individual holdings at downlands. The watershed approach helps in an integrated development of different parts of the watershed in accordance with their nature, problems and potential with best amount of interference. The integrated watershed management principles have accordingly been adopted for holistic development of drought prone and rainfed areas. In areas where there is no defined drainage system and internal drainage is choked in sand dune complex, the 'Index Catchment Approach' was considered for taking development activities. Index catchment is demarcated by considering the crest line of sand dunes or eroded rocky outcrops from the line of water divide (Vangani *et al.*, 1998). Focused efforts on watershed management were started after 1983 when Govt. of India (GOI) launched model watershed projects with involvement of Indian Council of Agricultural Research. Under this program 47 model watersheds were identified in different agro-climatic zones all over the country. Rajasthan got its first model watershed during 1986-87 as Jhanwar model watershed near Jodhpur. Later on, to cover large areas of dryland, the National Watershed Development Program for Rainfed Area (NWDPRA) was launched in 1990-91 by GOI. This scheme covered blocks with less than 30% area under irrigation. Under this scheme Rajasthan initiated development in 204 watersheds in 190 Panchayat Samities. To involve large segments of the rural community in this venture, new guidelines for watershed development were adopted from 1st April 1995, and subsequently revised in August 2001. In 2008 National Rainfed Area Authority issued common guidelines for watershed management that became effective from 1st April 2008.

Lessons learnt in watershed management

In Rajasthan out of 26.5 million ha (rainfed) requiring natural resource management through watershed development and alternative income generation activities so far only 4.6 million ha has been treated under various schemes. Area available for watershed treatment in western Rajasthan is 2.0 million ha, whereas so far only 0.274 million ha has been treated under various schemes (Govt of Rajasthan, 2006). One of the examples of the benefits of watershed development in degraded arid fringe is in the degraded Aravallis at Siha, Rewari (in Haryana), where the number of electrified wells have increased from 67 to 205, sprinkler set from 2 to 70 and irrigated area from 260 to 420 ha.

The livestock population increased from 882 to 1396. Consequently, milk production increased from 2997 to 5724 liters day-1 and overall income increased by 400 to 500% (Singh *et al.*, 1996). With proper soil and water conservation measures in watershed in the Aravallis, soil loss was reduced from 150 t ha-1 to less than 5 t ha-1. Development of water resources resulted in the increase of net irrigated area to 28 ha and cropping intensity from 128 to 210%. Food grain and fodder production also increased substantially (Bhardwaj and Dogra, 1997). Likewise, adoption of graded bunds, gully control structures, contour cultivation, intercropping, use of cover crop in rotation along with other improved package of practices proved sustainable in the semi-arid region of Rajasthan. Graded bund has reduced the run-off from 20 to 4.8% and soil loss from 24 to 4.12 t ha-1 y-1. Besides other benefits, intercropping on contour resulted in 48% higher grain yield (Singh *et al.*, 1997). By adoption of various development activities in Osian index catchment during a five-year period, cropping intensity increased by 31.4% and forage yield by 1.97 t ha-1. Construction of water harvesting structures helped to increase the groundwater recharge as indicated by rise in static water level. Sediment deposition against loose stone check dam was 3.86 m/ha/yr (Vangani *et al.*, 1998).

For *in-situ* rainwater management circular micro-catchment of 5% inward slope with LDPE lining was successfully demonstrated in watershed area for establishment of *Ber* and other trees. For severely eroded and gullied catchment loose stone check dams (LSCD) were constructed at 1 m V.I. in Jhanwar watershed on 17 gullies, which proved very effective in controlling further extension of gullies and stabilized all the gullies (Goyal *et al.,* 2007). For channel treatment three masonry anicuts and two loose stone anicuts were constructed on the main streams at Jhanwar and Baorali- Bambore watersheds, respectively, which resulted in substantial reduction in water velocity and erosion downstream. Temporary inundation upstream helped in regeneration of vegetation beside recharging of groundwater. In Sar watershed, artificial recharge of groundwater was superimposed in a 2.8 ha m pond with three infiltration wells to improve water availability for conjunctive uses. For moisture conservation soil, straw and plastic mulch were tried in Baorali-Bambore watershed. The grain yield of pearl millet was 32.67% and 28.12% higher for plastic and straw mulch, respectively, over no mulch.

In alternative land use system various systems like agro-horticulture with pearl millet/mung/moth + Ber/Aonla/pomegranate, silvi-pastoral system with *C. ciliaris* + *Prosopis cineraria/ Colophospermum mopane/Harwickia binata* were successfully established in watershed areas. Ditch-cum-mound fencing and cut and carry system was adopted for pasture development in Jhanwar watershed (Bhati and Goyal, 1997). For wastelands alternative crops like *Cassia angustifolia* and *Lawsonia alba* were successfully raised at appropriate locations in the watershed. For arable farming improved varieties of pearl millet, clusterbean, mung bean, moth bean, etc., were introduced.

Water Demand under Global Warming

As per general circulation models (GCMs) with different scenarios of greenhouse gas emission, the globally averaged surface temperature is projected to increase by 1.4 to 5.8oC between 1990 and 2100. The possible change in temperature due to global warming will have serious implications for the present ecosystem. The greatest threat will be increase in evaporative losses and water demands caused by higher temperature (Minitzer, 1993). Globally evapotranspiration trends are projected for +5% to +10% increase due to increased temperature by +2°C to +5°C under equivalent doubling of atmospheric CO_2 from pre-industrial level (Schneider *et al.,* 1990). Wetherald and Manabe (1981) found that global evaporation changes by 3% when temperature changes by 1°C. Similarly, Budyko (1982) suggested 5% increase in evapotranspiration demand for each degree Celsius rise in temperature. Enhanced evapotranspiration would be primarily a consequence of higher air and land surface temperature.

Even in tropics, where temperature increases are expected to be smaller than elsewhere, the increased rate of moisture loss from plants and soil could be considerable (Rind *et al.*, 1989; Parry, 1990). Lal and Chander (1993) suggested a rise in annual mean surface temperature by 2.0-3.5°C over the Indian subcontinent by the year 2090. According to them warming would be most pronounced over northwestern India. The normal average annual evapotranspiration of Rajasthan is estimated as 1701 mm.

A small increase of 1% in temperature (£0.42°C based on normal maximum temperature of Rajasthan) will enhance the evapotranspiration demand by 11.7 mm on annual basis. It will cause an additional annual water demand of 718 mcm and 2250 mcm for the whole state based on net irrigated area (61345 Km2) and total cropped area (192302 Km2), respectively (Goyal, 2004). Total available utilizable groundwater for Rajasthan is 11159 mcm and the increase of 1% in temperature will put additional stress of 6.43% to 20.16% on existing groundwater resources and will reduce the number of safe districts from 6 to 3, bringing additional districts in the category of 'critical' and 'overexploited'. An increase in temperature by 2-3% from normal (i.e. 0.82-1.24oC) will leave only 1 district in the category of 'safe' zone. The remaining 31 districts will be mostly in the category of 'overexploited'. The satellite data of Rajasthan shows a total wetland area of 3450 Km2 which includes 1239 Km2 ha as natural and 2210 Km2 as man-made. Increase in evaporation due to global warming will cause additional annual water loss of 40.4, 80.7 and 121 mcm for 1, 2 and 3% increase in temperature, respectively. Globally, projected increase in temperature/evapotranspiration demand is coupled with increase in precipitation by almost same magnitude. However, it is projected that shift will be more towards extreme events. The area with higher rainfall will likely have more rainfall and areas with less rainfall will have lesser rainfall.

Since this state is not blessed with good perennial river systems, any increase in water demand requires careful planning for future water resource development. More emphasis is needed on development of technologies for reducing water losses in reservoirs, conservation of rainwater and development of such crop varieties that require less water.

Water Resources of Ladakh

In India the risk-prone high altitude areas include the easternmost trans- Himalayan part of Ladakh and Zanskar range (Jammu and Kashmir) and some parts of Lahoul-Spiti (Himachal Pradesh). Truly described as cold desert with low population density, Ladakh constitutes the easternmost trans- Himalayan part of Jammu and Kashmir State of India comprising of Leh and Kargil districts, bordering Pakistan and China. Leh covers an area of 82665 Km2 and 51358 ha of reporting area, situated along the valleys of the Indus River with an estimated population of 117232. Leh is India's highest district, and one of its most arid, coldest and sparsely populated parts (Humbert-Droz, 2004). The district is sandwiched between two Himalayan ranges, Zanskar in west and Ladakh in east. Across the eastern range are the great highlands of Leh district, Chang Thang, which extends along the south of Ladakh and west, over the Indian-Chinese border into Tibet. To the north of Leh district is the district of Kargil and beyond it is the Karakoram Range. The Indus River runs through Leh district from south to north, collecting the water of melting glaciers and straddling several high-lying villages along its banks. Geo-ecologically Leh district is divided into almost two halves- a cold arid desert with minimal precipitation in the northern half, and high altitude grassland (Chang Thang) with perennial grasses in the southern half. The cold desert region of Ladakh, lying at an elevation of 2600 to 5030 m above MSL, is characterized by low relative humidity (20-40%), low atmospheric pressure (493 mm Hg), low partial pressure of oxygen, high wind velocity (8-10 km hr-1), very low annual precipitation (80-300 mm), and subzero temperature (up to -40°C) during winter months. Intense sunlight, high evaporation rate, strong winds and fluctuating temperature (30 to -40°C) characterize the general climate. It is generally said that a man sitting in the sun with his feet in the shade can have sunstroke and frostbite at the same time. With sparse vegetation there is little moisture in the atmosphere. Rains are very rare, though it may even snow during July-August, the hottest months. Because of high mountains all round, the area remains landlocked to the outside world for nearly six months in a year.

Irrigation Technology

Despite the mountains being global storehouse of fresh water (in the form of glaciers), the communities residing in them have to struggle with a severe shortage of water, both potable, as well as for irrigation. Nowhere does this cause greater hardship for habitation and livelihood than in the cold desert regions of the Himalayas. The reasons vary from climatic to geographic to human-induced ones. Depleting water resources are manifest in the form of thinning/retreating glaciers, sedimentation of water harvesting structures, etc. In the Trans-Himalayan region, the main form of precipitation is snow, but barely 250 mm or less per year. The economy of these regions is based primarily on agriculture, but the cultivation period is very short as the land is snow-covered for about 6 months. Water for irrigation comes almost from glacial melt, transported over long distances through small channels that are locally called canals/*kuhls*. In view of the prevailing arid conditions in the district, agriculture is possible only where water for irrigation is available; much of irrigation in the district is through canals/khuls.

Irrigation technology in Ladakh was transferred from neighbouring regions (Osmaston, 1985). The farmers have partly transformed the barren semi-desert into green croplands through skilful irrigated cropping that goes back to at least Tenth Century A.D., when it is said to have been introduced by the scholar saint, Atisa (Bell, 1928). The major source of water in Ladakh and other similarly cold arid areas is the glacial melt water through streams. Other sources are the natural springs, which are generally the outlet point of various watersheds, and whose discharge is controlled by groundwater from various glacial-melt streams. More often the streams run some distance away from the cultivable land, or are incised deeply below it, so that a long canal is necessary to bring water to the desired place.

It is not easy to lay the canal avoiding the natural obstructions, or to maintain a uniform gradient. It requires considerable traditional expertise. The skillfully engineered work sometimes gives rise to an optical illusion that it is flowing up-hill. The canal bed is often made of very porous material, loose stones and boulders, so there is considerable loss from seepage. The bushes and trees growing around the channels confirm that these losses take place. The Stongde Gompa (in Zangskar valley of Ladakh) canal loses nine/tenths of the initial flow of 0.01 m3/sec before reaching the Gompa (Crook and Osmaston, 1994). Wherever it is possible, farmers have skilfully diverted the water through construction of long canals, some of them running for few kilometres traversing through rocky mountains.

These canals were constructed during early stage of the history of the region. The glacial melt waters form various rivulets called *kangs-chhu* (ice water) which join together to form a *togpo* (stream). Togpo water is discharged/diverted to a valley touching many villages, in which the main canal, called *ma-yur* (mother channel), is constructed. It is built along a mountainside that forms its retaining wall, and is lined with clay to hold the water. This may be termed the Ladakhi version of a dyke. Water from the *ma-yur* is further diverted into *yu-ra* (small canal), which irrigates the fields. The point from where togpo water is diverted into *ma-yur*, and *ma-yur* water into a *yu-ra* is called yurgo. Ska is the point from where *yu-ra* water is diverted to the field. Water in the ska is further guided through channels known as *snang*, which carry the water throughout the field. The cultivated fields are generally on terraces and irrigation is through gravity gradient. Therefore, immense skill is required of the farmer to maintain the gradient of the field. The gradient should be such that it is neither steep nor flat, rather a gradient that allows smooth flow of water with enough seepage into the soil with negligible erosion. The water distribution through a system of channels is quite complex with different sizes of channels distinguished by various names. There may be some regional differences in the phonetics of the names. Moreover, it is similar in whole of Ladakh (Crook and Osmaston, 1994).

Ponds

In some villages, water from the *togpo* through *ma-yur* is stored in rdzing (pond). The water from this is then diverted into a *yu-ra* that carries it to the field. Individual families rarely construct their own pond. Every year in the beginning of spit (spring), ponds are cleared of silts. The villagers collectively undertake the cleaning operation. Usually the silts are piled up into *lut-pung* (small mounds), on the edge of the pond to hold the water, as this silt is used to prepare manures it is stored in dry lavatories. Usually a pond gets water supply from the village stream through the *ma-yur*. In some villages like Phey, a-Yu and Ska-ra ponds are also supplied water from *chhumig* (spring water). This is possible because they are situated at a lower level, and because the springs in these villages are well endowed.

Irrigation Water Management

Irrigation water is a critical source of food and wealth in any high altitude cold desert. One might expect to find a complex politics of distribution and exploitation. Distribution is certainly complicated, but exploitation is rare (Gutschow, 1993). There are inevitable abuses and conflicts in the distribution of water. Villagers are continually being watched at the same time that they are watching out for their neighbours, thus creating a reciprocal check on activities. It is difficult to evade signs of cheating, a moist field when all the surrounding fields are dry. The combination of an effective social organization with technical expertise suited to the local environment has enabled the farmers of cold desert to convert a few patches of their semi-desert to a very intensive and highly productive agriculture. This adaptation by the Ladakh farmers to a high altitude cold desert is the foundation of human settlement in such an inhospitable environment. The farmers in large villages are divided into groups commonly known as *Schhu-cho,* and each group gets right to water according to the traditional distribution system. The groups and the distribution pattern of water are recorded in the land records, stored at the patwari's office, whose duty includes maintenance of an official register of these water rights or on a silk document in the village itself, generally referred to as *bandabas* by local people. The documentation suggests that these disputes could be taken to the court. While such cases are beginning to be witnessed in some rare cases, mostly villagers still prefer to negotiate their disputes at the village level itself. Distribution of water for irrigation is in accordance with the rotational system. The rotational system is largely determined by the village topography, total village acreage, relative exposure to sun, average temperature, size of glacier, soil type and seepage in the irrigation channel, among other factors (Gutschow, 1993). The amount of acreage and the number of irrigation channels in any given village seems to determine which kind of rotational scheme is used. If the village is small and the irrigation channels few, rotational system of water distribution is pegged to the household, rather than the channels. For example in Leh and Sakti, water distribution is arranged by channels. In Phey (small village of forty-three household), water distribution is pegged to the household. This is because in small villages it is fairly easy to arrange and anticipate whose field will get water according to the rotational system. In a village where distribution is arranged by channels, the field lying along the given channel is irrigated in order, which may belong to a large variety of households. These predetermined rotational schemes between households, are in accordance to *bandabas.*

Chhur-pon

The word *chhur-pon* means Lord of the water, derived from *chhu* meaning water and *spon* meaning Lord. Water supply to individual families for irrigation is supervised by a *chhur-pon*. The *chhur-pon* is an official selected by the villagers, who is in charge of water distribution for irrigation and is perhaps the most important functionary in this regard (Koshal, 2001). This functionary used to be selected by consensus in earlier times. Nowadays rotation system is prevalent in most villages. The office of the *chhur-pon* has been in existence ever since people can remember, except in villages where water is available in abundance. In some villages the word *chhur-pon* is also assigned as individual house name (in Ladakh every house has a name). This is a clear indication of the importance of *chhur-pon* in these villages. If water is available in abundance then a *chhur-pon* is not required, as in the case of Hunder village of Nubra block. Even in villages where there are *chhur-pons*, his task is reduced to a great extent when water is in abundance. If the scarcity of water in a village is acute, then more than one *chhur-pon* is required, even if the village is small in size. In return for their service a *chhur-pon* is given *so-nyom* (one man load of cereal crop) after the harvest but nowadays this has been taken over in monetary terms, and the amount differs among villages. The *chhur-pon* has thus quite difficult duties to perform. As a customary respect for water, while stopping or releasing water from *ma-yur* to the *yu-ra* the *chhur-pon* has to keep his *gon-cha* (long overcoat) at ankle length and can not tuck it into his belt as is often done while working.

It is expected of the *chhur-pon* to distribute water according to the rota system, and monitor the activities of other farmers. At the operational level he is not much seen in the picture. Members of the village monitor each other's activities, and this system is inbuilt, as all the members are aware of how much and at what time water has to be diverted into their *yu-ra*. The farmer receiving water first monitors that the water is not diverted earlier than the decided time, and the farmer who receives later, monitors that his turn for water is not delayed. The first watering of the field occurs soon after the *sa-ka* ceremony is performed. Sa-ka is a festive-auspicious occasion. The day for *sa-ka* is decided by the village *onpo* (astrologer). On this day, all the work concerning cultivation, including construction of *ma-yur, yu-ra*, their repair work, etc., as well as the order in which it is to be done, is decided, and appropriate persons from the village are selected for various jobs. Only after making all the arrangements the agricultural work at the village level gets started. The *sa-ka* is performed in each village, or sometimes a group of villages performs it jointly. Its time varies from valley to valley or even from one sub-valley to another. It was customary to celebrate *sa-ka* ceremony on *gya-pe ma-zying2* (the mother field of the king), who possessed land in every village. After the conquest of Ladakh by the general Zoravar Singh, *sa-ka* was held in the *mazying* (mother field) of the aristocratic family. In villages where there is no aristocratic mansion, it is performed in the main field of the monastery. Besides, each family performs *sa-ka* on their own *ma-zying*.

The amount of winter snow determines the anticipated supply of water during summer, which further determines whether marginal fields should be sown or not in spring. The weather in spring determines whether the irrigation water supply will start early or late, since cool cloudy weather delays the snowmelt. This affects the ploughing and sowing time and the sequence in which the crops are sown. The application of sufficient but not excessive irrigation water is one of the most important farmer's skills. Insufficient water will result in poor growth, reduction in yield and possible risk of salinization. Excess can be dangerous to newly germinated crop, leach nutrients and waste of water. To avoid this situation when the crops are tender, only the skilled persons, especially the elders, are involved in irrigation of the field (mostly the first two irrigations, after the seeds are sowed). Later, when the crops are established well in the field, the less experienced younger people provide irrigation. Fields are irrigated for the last time in autumn after the soil is ploughed when harvesting of crop has been done. This practice causes the water to freeze in the soil, and prevents it from being flow out. The water is available again when the spring thaw occurs, and the soil gets moist for ploughing.

Watering a field

First Watering (tha-chus): Once the field is irrigated it is left for 3-15 days depending on the quality of the soil. The moist earth is called ser (gold), more precisely *chhu-ser* (literally *chhu* means water) i.e. earth is gold when it is supplied with right quantity of water. Soil with the right moisture is best suited for sowing and is referred to as *ser- phar- tog* (gold is ready); if it is too damp, then it is *ser-lchin-te dug* (gold is heavy). A snowfall in winter is a source of great satisfaction, for when it melts the earth absorbs the water and remains moist till ploughing and sowing begins. This kind of moist soil is called *kha-ser* (earth made gold by the snow). This saves farmers from first watering of the field and also the drudgery of waiting for their turn.

Second Watering (dol-chhu): After the field has been ploughed and sown, the second irrigation is given. Around 15-20 days of gap exist between the first and the second irrigation. The second irrigation is most important and is a delicate task because by this time the seeds have just sprouted: *dol* (to sprout), *chhu* (water), i.e. watering the tiny budding shoots. A skilled and knowledgeable person takes the responsibility of irrigating the field at this stage. Usually second irrigation needs more than one person. Insufficient or excess watering may lead the tender plants to get *tshik-ches* (burnt) or *shiches* (die). The *khyem* (spade) used for irrigation also serves as a measuring rod.

Third watering (sak-chhu): It is given 15-20 days after the second irrigation. It is generally a light irrigation, otherwise the crop will die. By this time the crop grows up to a height of about five inches (~13 cm). If denied irrigation at the right time, the upper portion of the crop dries up. In case someone's field is in great need of water, but his turn to irrigate is far away, then he requests someone irrigating the field to release one *sha-gugang- ngi* or *nang gang-ngi chhu* (a channel-full water).

Fourth Watering (non-chhu): It is given 10-15 days after the third watering. By this time the crop grows strong enough to stand a little excess or slightly inadequate supply of water. Anyone can look after the watering because it does not entail any specific knowledge. At the time of *non* (all the crop has grown to the same level) *chhu* (water) it is easier to find out which plants have withered away and which would survive. In other words, the condition of the field becomes clear. From now onwards the crop has to be watered regularly, about once a week depending on the moisture in the soil and weather condition. No special care is required for these watering except that the water is let in and then closed. These are generally undertaken by the less experienced younger people. These continue for about six times before the last watering, which is called *do-chhu.*

Last Watering (do-chhu): Application of *do-chhu* is very important. If the field is not watered now the ears of the crop get dry and start falling. *Dochhu* means watering of fully-grown crop field. The shoots are of equal height and ready to be reaped. By this time some members of the family water the crop for the last time, while other members are busy harvesting the *ol* (alfalfa).

Global Warming: Ladakh Perspective

The evidence of global warming is becoming abundantly clear with newly appearing reports on climate change. One of the major consequences of rising global temperature is that the glaciers and ice-shelves are receding. This is true for the cold desert of Ladakh also (Jogesh, 2008). With diminishing glacial resources and increasing variations in climatic and ecological conditions, impacts like flash floods and frequent locust attacks are increasing. The Leh's inhabitants are, therefore, trying to adjust to the new threats to their traditional livelihoods and sustenance. The Tsomoriri lake in the eastern Changthang valley is glacier fed so anything that happens to the glaciers will impact the lake. The locals say that there are signs to suggest that the water level of the lake has increased over the last couple of years. Trekking along the Tsomoriri wetlands to a small island in the lake is a place that was once a breeding ground for the highly endangered black-necked crane. Today, with rising water levels, the island is almost fully under water and the last crane was spotted here seven years ago. Only 60 black necked cranes exist in India and experts say, that if global temperature continues to rise, more than 100 million species will be at the risk of extinction. Ladakh appears to be at the starting point of the climate crisis in the region, which if continues, will threaten not just the biodiversity of the whole region but also most of its water resources (Jogesh, 2008). According to Behera and Rahul (2008) ecologically fragile geographical areas like Ladakh are most likely to be adversely impacted by climate changes, which can rightly be considered as global commons, often inhabited by traditional communities dependent on scarce natural resources for their livelihoods, which are often at subsistence or near-subsistence level. Hence, an understanding of these communities' perceptions and responses to climate change are essential to constructing effective sustainable development strategies that mitigate adverse impacts of climate change. They found that the perceptions of climate change are widespread and include observations of rise in temperature and heavy snow melt, less snow fall, heavy and untimely rainfall, biodiversity loss, etc. These changes have severely affected the use and management of natural resources and hence the livelihoods of the local people. Leh district receives about 20 mm rainfall in a year, which has risen to nearly ten times in the past few years (Sethi, 2007).

Evident more than 73% of the surveyed households in Leh district have perceived the increasing intensity of rainfall. This heavy and untimely rain in recent years has caused widespread damage to private and public property through landslides and flash floods. The flash flood of August 2006 in Leh destroyed homes, bridges and roads, cutting off villages from sources of communication and assistance, and inundated the fertile fields with rock debris and sand, rendering the land unsuitable for further cultivation. Inhabitants of Leh district feel that climatic change has affected their livelihood in some way. According to surveyed households despite a well functioning *Chhurpon* system, their villages face severe water scarcity, which is attributed to heavy runoff of water from faster snow melt, and erratic nature of sunshine hours causing unpredictable flow of glacial melt water into the streams. It is important to note that the run off water enters the Indus River and flows downstream. Ladakhis have traditionally not relied on the Indus for irrigation (except for few villages like Shey, Thiksey, Choglamsar, etc.) due to its shallow depth and difficulty in lifting water from its levels without the use of less efficient mechanized pumps that also pollute the environment. The altitude and climate of Ladakh pose major mechanical challenges that are often not experienced in any permanently inhabited part of the world (Behera *et al.*, 2008).

The disappearance of snow from the surrounding mountains has led to reduced grass yields important for grazing. As a result, the pastoralist groups now move more frequently from one place to other in search of grazing lands as compared to a decade earlier. This has further impacted the regenerative ability of the highlands. Another impact of climatic changes has been the increase in the frequency and intensity of pest attacks. The recent (2006) devastating invasion by locusts, both in magnitude and duration in Nyoma and Durbuk blocks is perceived by inhabitants of villages such as Anlay, Rongo, and Kuyul in these blocks as a consequence of the rise in temperatures in the region. It has been noted by the agricultural extension office and the Defence Institute of High Altitude Research (DRDO) at Leh that the frequency of such pest attacks can be attributed to warmer temperatures which is conducive to the increase in the geographical range and the duration of infestation of such agricultural pests.

Chapter 4

Climate Variability and Crop Production in Arid

Western Rajasthan

The hot arid region of India extends over an area of 317,090 km2, covering seven administrative states of India viz., Rajasthan, Gujarat, Haryana, Punjab, Maharashtra, Karnataka and Andhra Pradesh. Two states, Rajasthan and Gujarat, account for 81.5% of the total hot arid areas in India (Krishnan, 1968a, b). There are different methods of characterizing aridity of an area. In order to take up development work that would provide the largely agriculture-dependent inhabitants the capabilities to withstand impact of drought and other natural hazards in the drylands, including arid areas, Government of India followed the concept of demarcating arid and semi-arid areas on the basis of tehsil-wise water balance to determine moisture index. If the moisture index was below -66.6%, the area was considered arid and if it was between -33.3 to -66.6% it came under semi-arid zone. Based on this demarcation areas were identified for Desert Development Programme (in arid areas) and Drought Prone Area Programme (in semi-arid areas).

Climate and its Variability

Arid Rajasthan is characterized by limited seasonal precipitation with erratic distribution, high atmospheric temperature that has large diurnal and seasonal variation, strong insolation and persisting wind regime. Consequently, there are high crop water requirements. The weather conditions remain too dry, even in normal years, for most part of the year and are inhospitable for successful crop growth.

Rainfall and its Distribution

Normal dates of sowing rain (>25 mm week-1) in western Rajasthan are between 1st and 15th July, but the monsoon sometimes sets in as late as 3rd week of July in the eastern part and 1st week of August in the western part. Sowing rains have also been found to occur at times as early as 1st week of June in the eastern part and 2nd week of June in the western part. Such early or late commencement of sowing rains lead to large variability in yearto- year crop productivity. The assured crop growing period in western Rajasthan varies from 25 to 90 days under shallow soils and 35 to 105 days under deep soils (Rao *et al.*, 1994).

The mean annual rainfall in the region varies from 185 mm at Jaisalmer to more than 467 mm at Sikar. About 80-90% of the annual rainfall is received during the southwest monsoon when westsouthwesterly moving depressions are the main sources for *kharif* crops. The depressions originating from the Arabian Sea during May or early June move in east-northeasterly direction and also cause rainfall but such events occur with a lesser frequency of once in 2 or 3 years. The western disturbances during winter also cause precipitation, especially in the northern districts, which favour the *rabi* crops. The high inter-annual variability in rainfall is reflected in its high degree of coefficient of variation, ranging from 36% in the eastern part to 65% in the western part. This variability in annual rainfall is the most important factor, influencing the crop and pasture yields in the region.

Solar Radiation and Duration of Sunshine

Solar radiation is generally high throughout the year in western Rajasthan. During winter, it varies between 15.12 and 17.71 MJ m-2 day-1 and in summer months the values range from 22.79 to 26.50 MJ m-2 day-1, with a mean of 22 MJ m-2 day-1. The daily duration of bright sunshine hours in the area remains above 10 h day-1 in May and reduces to 6.6 h day-1 in July and August. In winter it is above 8.8 h day-1.

Air and Soil Temperatures

The region experiences extreme air and soil temperatures, which considerably increase the demand for water requirements by vegetation/ crops. During winter, mean maximum air temperature varies from 22.4°C to 29.0°C and minimum temperatures between 4.1°C and 14.3°C. Air temperatures sharply increase from April onwards and stands highest during May till pre-monsoon showers sets in the area. Summer air temperatures vary between 31.2°C and 42.0°C with peak values as high as 50.0°C. Temperatures fall during the monsoon period (June-

September), but rise by about 3.0°C to 5.0°C after the recession of the monsoon and again start falling up to -5.7°C from December onwards due to winter conditions. Soil temperatures follow the diurnal and annual cycles of air temperatures, and shoots up to 62.0°C during May and June, when in general, it is higher than air temperatures by 10.0°C (Ramakrishna *et al.*, 1990).

Relative Humidity

Relative humidity in the region is often less than 30% during summer months, but gradually increases to 80% by monsoon and then decreases from October onwards, following the withdrawal of the monsoon. Low humidity, combined with strong wind regime, leads to advection, a phenomenon that causes evaporation loss more than the energy actually available through solar radiation.

Wind Regime

The winds over the region blow from southwest during the monsoon and from northeast during the winter. Winds during the winter are low. From April onwards a strong wind regime builds up along with the increasing temperature, and reaches 15 to 18 km h-1. Peak winds occasionally reach as high as 60-80 km h-1 during severe dust storms. Strong wind regime during May and June cause wind erosion, depleting soil fertility.

Evapotranspiration Rates

The extreme climatic conditions in the region result in high evapotranspiration rates. The potential evapotranspiration rates are 1.3 to 3.5 mm day-1 during winter months of December and January. By May, the values rise to 3.7 to 9.5 mm day-1. During monsoon period the evapotranspiration rates are 5.2 to 10.6 mm day-1, but decrease from October onwards till December. The annual potential evapotranspiration ranges from 1400 mm in the eastern part to more than 2000 mm year-1 in the western part of arid western Rajasthan.

Weather Extremes

Frequency and Nature of Drought

The region is frequently prone to severe drought due to low and erratic rainfall. The frequency of agricultural drought at four locations of arid Rajasthan, as per Ramana Rao *et al.* (1981). Out of 108 years (1901-2008), the region experienced agricultural drought in one part or the other in 52 to 62 years, which suggest drought occurs in the region once in three years to alternate year.

Jaisalmer district is most prone to drought. During 1901-2008, the agricultural drought occurred here in 70% of the years, out of which 43% of the years experienced severe drought and 27% years moderate. Barmer district experienced severe drought in 30% years and moderate drought in 20% years. Bikaner experienced severe agricultural drought in 23% years and moderate in 25% years, whereas Jodhpur district experienced severe drought in 17% years and moderate drought in 28% years. Often, drought persists continuously for 2 to 6 years. Prolonged drought was experienced in the region during the decades 1901-10, 1911-20, 1932-40, 1961-70, 1981-90 and in 2001-2006, with 1918 and 2002 standing out as the worst drought years during the recorded period. The moderate to severe drought frequency varies from 1 out of 5 years in Churu district to 2 to 3 out of 5 years in Barmer and Jaisalmer districts.

Probability of occurrence of early, mid and late season drought The weekly rainfall pattern at Jodhpur during 1971 to 2008 indicated that the rainfall distribution was fairly good in 17 years, whereas in the remaining 21 years, the region experienced drought at one stage or the other. Further classification of these 21 years' drought, based on weekly rainfall distribution, showed that the probability of early season drought was 26% (9 years), mid-season drought in 9% (3 years) and late season drought in 26% (9 years). Among the three categories of drought, the impact on grain yield of pearl millet and pulses was felt more under mid-season and late-season drought. If the drought is severe at early season of the crop and reduction in rainfall in subsequent crop stages, the impact of such severe drought is felt in the region on food, fodder and drinking water.

Impact of drought on livestock population

The region supports a large livestock population, out-numbering humans by about 4 times in Jaisalmer district, 3 times in Barmer, 2 times in Bikaner and 1.5 times in Banaskantha, Jalor and Jodhpur districts, as against only one-half in the rest of the country. Ahuja (1994) showed that the supply of forage was 37% shorter than the demand at the current level of livestock population. During severe drought years, the normal forage productivity was declined by 82% in rainfall zone of <200 mm, 74% in 200 to 300 mm rainfall zone and 42% in 300 to 500 mm rainfall zone (Shankarnarayan and Singh. 1990). Consequently, during drought years, the livestock population decreases in the Indian arid zone due to scarcity of fodder and water. Sheep, goat, buffalo and cows were the worst affected during drought years. The consecutive drought during 1985, 1986 and 1987 resulted in decrease of livestock population (26% in cattle, 7% in sheep; Source: DAH, 2005).

Flood Scenario

Floods in this arid region are rare as compared to drought. The decadal-wise frequency of the flood-causing rainfall (called flood years), however, indicated that it occurs with a frequency of 2 to 3 out of 10 years. The decades 1911-1920, 1941-1950 and 1981-1990 were notable in this respect. Broadly such rainfall has been recorded in alternate decades, e.g., in 1970-1979 and in 1990-1999. Widespread flood-causing rainfall was witnessed during 1908, 1917, 1944, 1975, 1983 and 1997.

During

1917, the average rainfall in the region was exceeded by +161%. Such widespread excess rainfall was not recorded in subsequent years (Rao *et al.*, 2006).

Cold and Heat Wave Conditions

Extreme temperatures are frequently encountered in arid Rajasthan, which cause great threats to the growth and productivity of agricultural crops in both *kharif* and *rabi* seasons. The cold or heat wave conditions many times follow in a sequence, affecting crops at various stages of growth. Often air temperature falls below -4.4°C in winter, resulting in frost injuries in *rabi* crops like mustard, castor, cumin, gram, barley, taramira and vegetables. In association with passing western disturbances, severe cold wave conditions prevail during December and January. The districts of Churu, Jhunjhunun, Ganganagar, Hanumangarh, Sikar, Jaisalmer and Bikaner are more affected by intense cold wave conditions.

Climate Change in Arid Rajasthan

The impact of projected climate change by the end of 21st Century (IPCC, 2007) is more likely in arid region than in the semi-arid or sub-humid regions of India. The arid phase of northwest India has a history of about 3000 years (Pant and Maliekel, 1987). In the northwest India, covering Punjab, Haryana, west Rajasthan and west Madhya Pradesh, there was a marginal increase in the rainfall by 141 mm and fall in air temperature by - 0.52°C in the past 100 years (Pant and Hingane 1988; Rao and Miyazaki, 1997). The secular trends in annual rainfall in the arid western Rajasthan showed no significant change during 1901 to 2008.

To know the impact of introduction of large-scale irrigation through *Indira Gandhi Nahar Parjyojana* (IGNP) canal into the region, past hundred years of annual rainfall and air temperature data were analyzed from selected locations along the tracts of the canal (Rao, 1996, 2005). Long-term annual rainfall at Ganganagar showed an overall increasing trend at the rate of 1.18 mm year-1. Even during drought years, Ganganagar area received higher rainfall than the other arid areas, attributed to the irrigation imposed for longer periods that supplied the moisture to rainfall systems and also indirectly through energy balance from increased vegetation. However, increase in rainfall was not observed in adjoining Bikaner and Jaisalmer districts, where the irrigation was introduced since past 1-3 decades.

Strategies to Cope up Weather Extremes and Climate Change

Early Warning of Drought and Crop/Forage Yields

Simulation of crop production from arid regions had several difficulties due to temporal and spatial variability in rainfall, high evapotranspiration rates and heterogeneity in land use or crop acreage. Because of our limited knowledge in understanding the assimilation of carbon by various arid zone crops, by and large we have to depend on regression models or soil moisture simulation models for explaining the crop/forage yield.

For drought and crop surveillance over the region, computer models like the SPAW (Soil-Plant-Air-Water) and CERES-millet have been tested and modified at CAZRI. The SPAW model was used for assessment of soil moisture and crop stress conditions under pearl millet, which could explain 89% of pearl millet yield in Jodhpur district.

The relationship between pearl millet grain yield (kg/ha) and water stress index (WSI) of SPAW model was Y = -45.38 WSI + 526.18 (r=0.9427) (Rao and Saxton, 1995; Rao *et al.*, 2000). Of the various crop growth simulation models developed by the International Benchmark Sites Network for Agrotechnology Transfer (IBSNAT), using the DSSAT (Decision Support System for Agrotechnology Transfer), the CERES-millet model was used for simulation of pearl millet (Ramakrishna *et al.*, 1994), which predicted grain yield and other parameters closer to the observed yield.

RANGETEK model was used for prediction of forage yield of promising arid grasses like *Lasiurus sindicus, Cenchrus ciliarias, C. setigerus* (Singh *et al.*, 1996a, b). The model was used for real-time simulation of daily soil water content, soil evaporation and plant transpiration to forecast forage production in terms of a ratio of actual to potential transpiration (yield index). The deviations between recorded and estimated forage yield using the model were from 8 to -12% (Singh *et al.*, 1996a). The studies on *C. ciliaris* revealed that the predicted yields using the RANGETEK model were between -10% and +15% of actual forage records (Singh *et al.*, 1996b). These 1-dimension model combined with a knowledge based geo-spatial Decision Support System (DSS) could be an effective tool for drought early warning and management by farmers, experts and general end-users.

Evapotranspiration Rates and Water-use Efficiency of Crops

Evapotranspiration requirements of crops were measured in gravimetric lysimeters, keeping the crop under unstressed conditions (100%PET rates of irrigation). The crop coefficients at different growth stages, water-use efficiency (kg/ha/mm) were computed for planning scarce water resources of arid area for increasing the productivity per unit of water.

Rainwater Harvesting for Coping with Drought

Rainwater harvesting is an old practice in arid Rajasthan. Both *insitu* as well as *ex-situ* techniques can be utilized to harvest runoff rainwater for productive uses like drinking and supplementary irrigation purposes. It is estimated that nearly 300 mcm of water for human consumption, 350 mcm for livestock and 6800 mcm for irrigation is required by 2010 (Venkateswarlu *et al.*, 1990).

Crop Contingency Plan for Delayed Monsoon Condition

CAZRI (Narain *et al.*, 2006) prepared contingency plans for delayed monsoon. Delay in monsoon reduces the water availability period and need strategies to grow crops, which match and yield better under such conditions.

Trends in Arid Zone Research in India

Such contingency plans were found useful to obtain better economic returns compared to traditional systems.

Use of weather based Agro-Advisory Service

Weather based agro-advisory bulletins to farmers provide information on a real-time basis so that they can avail advantage of climatic information to minimize crop losses caused by unfavourable weather. Since 1998 CAZRI is issuing bi-weekly agro-advisory bulletins based on Medium Range Weather Forecasts prepared by National Centre for Medium Range Weather Forecast (NCMRWF). The information on the type of weather situations likely to be encountered and the method to be adopted for efficient management of inputs, is provided to the farmers in advance for minimizing weather related risks.

To measure the economic impact of agro-advisory services (AAS), a three-year survey in Manai, Narva and Palri Mangalia villages of Jodhpur district was conducted during 2003 to 2006. The economic analysis on benefit-cost ratio for *kharif* and *rabi* crops indicated that for pearl millet the benefit-cost (B:C) ratio was 2.15 to 3.17 for AAS farmers, whereas for non- AAS farmers it was 0.73 to 2.80. In case of mustard, farmers were much benefited as the B:C ratio for AAS was 1.92 to 2.25 against 1.64 to 2.13 for non-AAS farmers (Rao, 2008).

Land Use of Arid Lands in India

Land use is the human modification of natural environment of wilderness into built environment. Land cover is the observed bio-physical cover on the earth's surface while land use is the arrangement, activities, and inputs people undertake in a certain land cover type to produce, change or maintain it (FAO, 2005). India has about 18% of world's human and 15% of livestock population to be supported with only 2.4% of the world's geographical area and 1.5% of forest and pasture land. Per capita availability of land in the country has reduced from 0.4 ha in 1950-51 to 0.14 ha in 2000- 01. With the prevailing trend it may be further reduced to 0.077 ha by 2020.

In the hot arid ecosystem of India low and erratic rainfall, extremes of temperature, high evaporation loss, saline and meagre groundwater, absence of perennial streams, and dune-covered and rocky/gravelly terrain, are the major factor influencing the land uses. Agriculture, which is the most dominant land use, is mainly rainfed and subjected to high risk and uncertainty. The cold arid Ladakh region of Jammu and Kashmir state is situated in the high Himalayan Mountain, where arable land is very limited and the length of growing period (LGP) is short. Human activities are very limited and so are the livelihood sources.

Before the 1960s land use studies were primarily based on revenue and land records data. Thereafter the use of topographical maps and aerial photographs made it possible to show the spatial distribution of different land use systems. After the 1970s satellite remote sensing and GIS virtually revolutionized the research on land use mapping and monitoring of changes over time and space. Provides a short review of the major findings on land use studies on arid ecosystem of India.

Land Use: A Historical Perspective

History of land use and land classification system in the arid ecosystem is reported by Roychaudhury (1966, 1970). Evolution of land use in north-west arid zone of India has been reported by Whyte (1960),

Bharadwaj (1965) and Randhawa (1980). According to these studies land use in the region evolved through the Neolithic Period when man had taken up sedentary life and had domesticated the animals.

They first settled along the rivers and near water points, and later on shifted to other marginal areas. James Todd's Annals and Antiquities of Rajasthan (1929; reprinted 1978) and Erskine's Western Rajputana States Residency and Bikaner Agency Gazetteer (1909) highlight the land use system that prevailed during British Period but no systematic account is given. Classical treatise of Nainsi Mohto (Singh, 1969; Jain, 1974) on *Marwar-ra Pargana-ri Vigat* provides an account of cultivated area and crops being produced in the 'Khalsa' villages which were directly administered by the rulers. No account was available for villages under the *Jagirdars*.

After the settlement of land holdings, abolition of *Jamindari* system and reorganization of states, the use of each land holding, including the area under different crops as per cadastral map of a village began to be recorded in a register called the *Girdawari*, and it continues till today. Kanitkar (1952) estimated that out of 88597 miles2 area in Rajputana desert only 20694 miles2 was available for cultivation, while net sown area was 8124 miles2. Irrigated area was nearly 1486 miles2. Banerji (1952) while defining the role of vegetation in desert control estimated 320 miles2 of *'Khalsa'* area under forest department. Total *khalsa* area was 5,545 miles2.

Thus forest covered only 5.8% area. Culturable waste covered 523 miles2. Nearly 31,000 miles2 was with the *Jagirdars*.

Land Use Classification System

The Land Records Office at tehsil or taluka level keeps land use record of each village under their jurisdiction using a 44-column proforma. This classification system categorizes fallow lands into three sub-classes under individual and government holdings; area irrigated by different sources and subclasses of rainfed, double cropped and net sown areas. Based on this format the State-level Directorate of Economics and Statistics compiles and publishes annually the land use data by adopting a 16-point format and 12 land use classes. Irrigated area is not included in this classification system, but is given separately under irrigation. At national level the land use data is published by the Directorate of Economics and Statistics, Department of Agriculture and Cooperation, under a nine-fold classification system. The district Census Handbook published by the Directorate of Census Operations also includes land use data in five classes, viz., forest, irrigated area, rainfed area, culturable waste and permanent pastures, and area not available for cultivation. The spatial pattern of the land uses was partly available from the large-scale topographical sheets prepared by Survey of India.

Systematic mapping of land uses in the arid zone began after the establishment of Central Arid Zone Research Institute in 1959, especially as a component of integrated basic resources survey and mapping. Taking into consideration various land use classification systems available then, the typical problems of arid zone and the feasibility of mapping, A.K. Sen evolved a land use classification system for the Indian arid zone, which is summarized in Sen (1978, 1982). Sen (1972a, I977a, b) also developed aerial photo interpretation techniques to classify and map different land use categories. The classification contains 15 land use categories. Cultivated lands were classified into seven categories based on the intensity of cultivation to avoid repetition of land use survey every year, since the annual extent of cropped area varies and depends on the intensity, distribution and amount of rainfall. Wastelands were classified into four categories viz., sandy, saline, rocky/ stony and rocky and gravelly waste, to which were added subsequently the gullied waste, mud waste and waterlogged areas. It formed a basis for the future nation-wide land use/land cover mapping from satellite data (Gautam and Narayan, 1988; NRSA, 1989) that identifies six land use/land cover categories under Level-I and 24 categories under Level-II, as well as the wasteland mapping units (NRSA, 1986, 1991). The satellite-based classification system was further refined during the Integrated Mission for Sustainable Development (IMSD) project of ISRO, when classification up to Level-III was possible (i.e., kharif and rabi croplands; dense and open forest/grassland cover; NRSA, 1995a; Balak Ram, 2002). The classification system has further evolved in recent years with the availability of higherresolution satellite images.

Present Land Use

As per land use statistics of 2005-06 out of 42.5 mha area of hot arid ecosystem, cultivated lands constitute 63.67%. This includes 51.69% net sown area, 6.26% current fallow and 5.72% other fallow land. Net and gross irrigated areas are 17.26% and 26.75%, respectively. Double cropped area is 14.09%. Among the other uses, forestland occupies 5.72%, pasture 3.21%, land put to non-agricultural uses 5.3%, barren and uncultivable land 9.10% and miscellaneous tree crops and groves 0.18%. Land use statistics of the hot arid zones of Rajasthan, Gujarat, Punjab, Haryana, Andhra Pradesh and Karnataka as per revenue. Satellite-based mapping has revealed that the region has 9.12

mha or 21.56% area under wastelands (NRSA, 2005) of different categories. Arid western Rajasthan has its 29.4% area under wastelands, Andhra Pradesh 22.27%, Gujarat 13.17%, Karnataka 11.22%, Haryana 3.58% and Punjab 4.07%. Sandy wastes are most dominant in Rajasthan, barren rocky/stony waste and salt affected lands in Gujarat, and degraded forest and land with/without scrub in Andhra Pradesh and Karnataka. Pearl millet, cluster bean, moth, mustard and wheat are the important crops in arid Rajasthan, cotton, wheat, rice and mustard in Haryana and Punjab, groundnut, cotton and castor in Gujarat and green gram, red gram, groundnut and sunflower in Andhra Pradesh and Karnataka.

In the cold arid region, out of 9.67 mha total area (in Leh and Kargil districts) the reporting area is only 64,626 ha. Rest of the area is covered by snow and hill ranges. Out of the total reporting area cultivated lands constitute 30.92%, which includes 29.67% net sown area, 0.53% current fallow and 0.72% other fallow land. Forest occupies 0.1% area, land put to non-agricultural uses 6.32%, barren and uncultivable land 45.03%, pasture 1.69%, miscellaneous tree crops and groves 2.49% and culturable waste 12.36%. Wasteland statistics of NRSA (2005c) revealed 5.06 mha (88.28% of the total geographical area) under wastelands in the two districts. This includes 52.6% barren rocky/stony waste, 12.05% steep sloping area, 30.15% snow covered area, 2.68% sands and 0.18% degraded pastures. Ragi, wheat, barley, millet and apricot are important crop of this region.

Land Use Changes

Between 1982-83 and 2005-06 net irrigated and double cropped area in hot arid region has increased by 128% and 70%, respectively. Net sown area has increased by 2.4% only, current fallow by 1.7% and other fallow land by 12.1%. More impetus has been given to irrigated cash crops. The area under forest has increased by 4.6% and land put to non-agricultural uses by 16.6%. On the other hand, the area under culturable waste has declined by 18.6%, pasture by 7.7%, miscellaneous tree crops and groves by 36.8% and barren and uncultivable land by 2.2%.

Arid western Rajasthan has recorded substantial increase of 12% in net sown area, while forest has increased by 67.4%. Area under culturable waste has declined by 26.3%. This shows that though marginal lands have been brought under cultivation, the intensity of cultivation, particularly in rainfed lands, has not increased to that extent. This has increased the fallow lands. Environmentally sensitive common lands have become more victims of degradation by nature and man. Uneconomic use and over-exploitation of natural resources have caused problems like secondary salinization, waterlogging, mined spoil areas, degradation of groundwater, intensification of sand movement, blockade of natural drainage and spread of industrial effluents. Silting and pollution of irrigation tanks/ponds, degradation of mangroves and reduction in biomass are the other impacts.

In the cold arid region, horizontal and vertical development of agricultural is not up to the mark. Between 1982-83 and 2005-06 the net sown area has increased by 5.1%, other fallow land by 151.1%, land put to non-agricultural uses by 37.23%, and double cropped area by 86.3%. On the other hand the area under culturable waste has declined by 19.16%, miscellaneous tree crops and groves by 45.9% and current fallow by 52.77%. As per forest department there is 13166 km^2 forestland and 13130 km^2 area under wildlife conservation.

In western Rajasthan, Directorate of Agriculture, Government of Rajasthan occasionally brings out district-level trends in land use (e.g., from 1956-57 to 1990-91; Anon., 1992b). Based on field mapping, Balak Ram (1988) provided a quantitative account of land use changes in Siwana region between 1960-61 and 1985-86 at village, tehsil and regional level. Since then, several studies have been carried out on land use change mapping. Notably, Balak Ram and Rastogi (2004) mapped the land use changes in Kachchh district, while such mapping has recently been completed in Jaisalmer district (Balak Ram and Chauhan, 2009). Balak Ram and Gupta (2003) assessed the impact of industrial effluents on land use changes along the Jojri, the Bandi and the Luni rivers. Cincotta and Pangare (1991) discussed agricultural land use changes and pastoralism in Gujarat, while Chaurasia *et al.* (1996) discussed land use change in Punjab. Nautiyal *et al.* (2005) described ecological impacts of land use intensification in the Himalayas. Analyses of the earlier trends are summarized in Balak Ram and Gheesa Lal (1997, 1998) and Balak Ram (2003).

Issues and Challenges

Important issue and challenges of hot arid region are risky and traditionally subsistence rainfed farming with low productivity, degraded grazing and forest land, traditional and low productive animal husbandry, mined spoil areas, over-exploitation and injudicious use of ground water, land degradation and drought. In cold arid region the important issues are soil and water erosion, traditional use and non-managed pasture and forest land, subsistence and segregated farming system, poor system of animal husbandry and livestock production, neglected human and cultural value, under-utilized and ill-managed rain and river water. These are being constantly addressed by different government and non-government programmes through numerous institutions of research and development, but the vastness of the country, vagaries of climate and very high population pressure make the impacts manifest either slowly to, or are reversed fast.

Variability in Arid Soil

Characteristics Arid region in India is spread over in 38.7 million hectare area. Out of the total, 31.7 m ha lies in hot and remaining 7 m ha lies in cold region. The hot arid region occupies major part of northwestern India (28.7 m ha) and remaining 3.13 m ha is in southern India. The northwestern arid region occurs between 22030' and 32005'N latitude and 68005' to 75045' E, covering western part of Rajasthan, northwestern Gujarat and south-western parts of Haryana and Punjab.

In southern India it occupies parts of Andhra Pradesh, Karnataka and Maharashtra. About 62% area of arid region falls in western Rajasthan followed by 20% in Gujarat and 7% in Haryana and Punjab. Andhra Pradesh, Karnataka and Maharashtra together constitute about 11% area of arid region. The boundary between arid and semi arid region in Rajasthan cuts across the Jalore and Pali districts, goes roughly along the boundaries of Nagaur and Ajmer. The arid zone extends to Mahendragarh and Hissar districts of Haryana; Sangrur, Bhatinda and Firozepur districts of Punjab; Jamnagar, Kachchh, Banaskantha, Mehsana, Surendra Nagar and Rajkot districts of Gujarat.

The arid districts of Haryana and Punjab exactly border to north of western Rajasthan, while districts of Gujarat touch the southern boundary. Thus arid zone in India forms a contiguous tract, keeping western Rajasthan in the centre. In the arid regions of Rajasthan, Haryana, Punjab and eastern Gujarat, aeolian and alluvium are the major formations. These aeolian and alluvial parent materials are the quaternary formations. During that period this region witnessed wide spread alluvial sedimentation. However, the overall environment was semi arid (Dhir, 1989; Dhir *et al.*, 1994; Gupta *et al.,* 2000). This alluvium was reworked by the subsequent activities. Since the last glacial period the region has experienced two major periods of aridity separated by a long period of amelioration of climate during which natural vegetation reached its maximum extent (Singh, 1977). The Dune forming aeolian activities began at least 200K years ago and terminated around 13,000 years B.P. (Singhivi *et al.*, 1983). The vast alluvial plain of Haryana and Punjab was formed out of sediments deposited by the rivers flowing from the Shivaliks. During arid phase this alluvium was reworked by aeolian activity and resulted in the formation of sand dunes and sand plains (Sidhu and Sehgal, 1978).

The cold arid region occupies 7.0 m ha area covering Ladakh, Kargil districts in Jammu and Kashmir and Lahul Spiti district in Himachal Pradesh. The region is characterized with mild summer and annual rainfall of less than 150 mm. The soils occur on summits and ridge tops, mountain and valley glacier, glacio-fluvial valleys and fluvial valleys of greater Himalayas. Soils are skeletal and calcareous with alkaline in reaction and low to medium in organic matter content (Pratap Narain, 2008).

The Soils

. The soils of this tract have been mapped in Entisols, Aridisols and Alfisols soil orders. The Entisols cover maximum 17134.26 thousand hectares (54.21%), followed by Aridisols and Alfisols. The area of latter two is 14254.32 and 213.10 thousand hectares, comprising 45.1 and 0.67% of hot arid India. Haplocalcids, Haplocambids, Haplosalids, Petrocalcids and Haplogypsids constitute Aridisols at the great group level, while Torripsamments, Torriorthents and Torrifluents are the part and parcel of Entisols. The Natrargids, Pleargids and Haplargids are the great groups mapped in Alfisols. The spatial distribution of these great groups. Haplocambids dominantly occur in the arid part of Haryana, Gujarat and Punjab, covering around 77, 42 and 39.6% of delineated area in their respective state. Torripsamments dominantly mapped in 47.1% area of western Rajasthan, while the share of these soils to the marked area is around 43,14.3 and 9.7% in Punjab, Haryana and Gujarat, respectively.

Haplocalcids are the second dominant soils of Gujarat after Haplocambids with 11.1% of allocated area under arid land. Petrocalcds and Haplocalcids together constitute 8.4% area of western Rajasthan. The contribution of Haplosalids to the delineated arid land is higher in Gujarat than Rajasthan. Torriorthents, the shallow skeletal soils associated with hills are mapped in 7.8 and 4.3% area of arid part of Rajasthan and Gujarat, respectively. Torrifluents fine textured deep soils along the course of river constitute around 2% area of arid zone both in Rajasthan and Gujarat. Haplogypsids and Haploargids represent relict soils in Rajasthan and Gujarat, while Natrargids and Pleargids are the paleosols mapable only in the latter (Shyampura *et al.*, 2002).

The soils on summits and ridge tops of greater Himalayas are shallow to medium deep, excessively drained, sandy-skeletal to loamy skeletal, highly calcareous and severely eroded with low organic carbon and available water capacity (Sidhu *et al.*, 2007). While, the soils of

Mountain and valley glacier are similar in characteristics except they are covered with snow for a longer period. These soils were classified as Lithic Cryorthents and Typic Cryorthents. The glacio-fluvial valleys and fluvial valleys soils of greater Himalayas are medium deep to deep, well to excessively drained, loamy skeletal to coarse loamy and fine loamy soils with slightly acidic to neutral with medium to high contents of organic carbon. Some soils are highly calcareous and slightly to moderately alkaline have been classified as Typic Udorthents/ Udifluvents, and Dystric Eutrudepts great group, respectively (Sidhu *et al.*, 1997).

Physical Properties

Soils under Entisols are generally found in those parts of arid regions where high aeolian or fluvial activities are observed. The average soil depth of this type of soil is 118 cm. Average bulk density of these soils is 1.58 Mg m- 3. Surface horizon is rich in sand content in comparison to subsurface horizons. Soil texture is sandy with average sand content of 89%. In the extreme arid part at Jaisalmer, Bikaner, and Barmer sand content is > 90% for most of the places. Soil colour is mainly yellowish brown. The hue of the soil is 10YR, value ranges from 5-6, and chroma ranges from 3-4. Surface as well as subsurface soils are single fine grain in structure. Soils are mainly loose, non-sticky and non plastic in consistency. Most of the soils are well drained and fall under very rapid (>25.4 cm hr-1) class of O'Neals permeability class. Water retention at 1/3 bar (field capacity) on an average is 6.69% (0.07 cm-3 cm-3 or 70 mm ha-1 of soil depth). Similarly, the water retention at 15 bar (permanent wilting point) is also very low (2.62%) In contrast to Entisols, Aridisols are dominant in those parts of arid regions where aeolian activity is comparatively less. The average soil depth of this type of soil is 100 cm with well demarcated horizons. Concretions of calcite below the soil profile are a common feature of these soils. Average bulk density of these soils is 1.43 Mg m-3. Average sand content of Aridisols is lesser than Entisols and is around 70%. Soil colour is mainly yellowish brown to dark brown or pale brown. The hue of the soil is 10YR, value ranges from 5-6, and chroma ranges from 4-5. Surface as well as subsurface soils are single fine grain in structure. At few places, fine weak subangular blocky structure is found at subsurface soil layers. Soils are mainly loose, non-sticky and non plastic in consistency. At few places, presence of high clay and organic carbon at subsurface layers make the soil massive, friable, and slightly sticky. Most of the soils are well drained and fall under either rapid (12.5-25.4 cm hr-1) or very rapid (>25.4 cm hr-1) class of O'Neals permeability class. Water retention at field capacity (FC) and permanent wilting point (PWP) in Aridisols are comparatively higher than Entisols and are 13% and 5%, respectively. Soils under Alfisols are found in few places of arid regions of India. In comparison to Entisols or Aridisols, these soils are comparatively rich in clay content (36%). Organic carbon content of Alfisols is also higher (3.7 g kg-1) than Entisols (1.5 g kg-1) and Aridisols (1 g kg-1). Average permeability of these soils are moderately slow. Due to presence of high amount of clay, water retention at both FC and PWP arc high (23% and 10%, respectively).

Nutrient Status of Arid Region Soils

Characteristically these soils are very low in organic matter/ humus and most of the nutrients reserve is present in un-weathered mineral forms. These soils have low clay and silt, and therefore nutrient adsorption and retention by these soils are very low. Soils are generally alkaline in nature and high in soluble salts and calcium contents. Organic matter and Nitrogen Arid zone soils are low in organic matter because of low vegetation cover, high temperature and coarse texture. Organic carbon content in soils below 300 mm rainfall zone ranges between 0.05-.2% in coarse textured soils 0.2-0.3% in medium textured and 0.3-0.4% in fine textured soils. Dhir (1977) has shown that even in low range of organic carbon (>0.5%) increase in clay content is associated with rise in organic carbon. With increase in clay content there is increase in organic carbon content. Joshi (1990a) has reported that bulk of organic carbon (67-89%) is present in non humic form followed by fulvic and humic form and mean value of HA-carbon/FA carbon increased with increase in silt+clay. Aggarwal and Lahiri (1977), Aggarwal *et al.*, (1993) reported build up of organic carbon and nitrogen and C:N ratio (10:1) under vegetation cover. C:N ratio ranged between 5.1 and 6.9 and 3.9 and 10.1, respectively, in surface and sub-surface soil. Nitrogen in soils is present as organic (95-98%) and inorganic but in arid region soils, the inorganic form constitutes about 5 to 18% of the total N. Aggarwal and Lahiri (1981) reported 46 to 67% N as inorganic or mineralized in unstabilized sand dunes of western Rajasthan. Lower organic carbon and available nitrogen have also been reported in the soils of aeolian plains than in other landforms. Total N content of arid soils varied widely, as coarse textured Cirai, Pal, etc. contained 0.028% to 0.050%, medium textured Pipar and Soila (0.042 to 0.056%) and fine textured Asop contained 0.060 to 0.065% N. Aggarwal *et al.*, 1975 reported relatively highest concentration amino acid bound N amongst the various forms in arid soils of Rajasthan.

Phosphorus

Phosphorus present in soils as organic and inorganic forms but organic form constitute hardly 10-20% of the total phosphorus. Total phosphorus content in soils ranges between 300-1500 μg g-1 (Mehta *et al.*, 1971; Deo and Ruhal, 1972; Choudhari *et al.*, 1979) and about 80% of the inorganic P remain bound with Ca. Al-P was higher, generally higher than Fe-P. Total and inorganic phosphorus were irregularly distributed in the soil profile of arid soils, but organic phosphorus decreased with depth (Pareek and Mathur, 1969; Talati *et al.*, 1975). 15-20% (97-110 kg ha-1) of the total P is present in organic form as phytin, lecithin, phospholipids and other unidentified compounds (Tarafdar *et al.*, 1989). The available phosphorus content varies widely in different soils and is about 2.4 to 3.9% of total P and the mean content in different soil series is less than 10 μg g-1 (Mathur *et al.*, 2006). Even though soils are often medium to low in available P, response of P-fertilization in arid soils is generally observed only in good rainfall years (Aggarwal *et al.*, 1989).

Microbial Mobilization of Soil Phosphorus

An extensive survey through arid zones revealed that the population of P-solubilizing micro-organisms is relatively low and varied from one place to place (Venkateswarlu *et al.*, 1984). Bacteria were the dominant group in all the soils followed by fungi. But isolation of actinomycetes especially Streptomyces spp. which can solubilise inorganic phosphates, from arid soil was reported (Rao *et al.*, 1982). However, P-solubilzers isolated from arid soils could not perform satisfactorily in the field because of lack of competition with the native flora. Organic P-compounds must be hydrolyzed by phosphatases of either plant or microbial origin, into plant available form.

Tarafdar *et al.* (1989) reported that activity of phosphatases in different arid soils varied with landuse pattern. High activity of acid and alkaline phosphatases were observed in the rhizosphere of legumes (Tarafdar and Rao, 1990). Tarafdar *et al.* (1988) isolated a number of phosphatase producing fungi (PPF) belonging to Penicillium and Aspergilus from arid soils and their field application enhanced dry matter production and grain yields of clusterbean and mung bean (Tarafdar *et al.*, 1992, 1995).

Arid soils contain 7 to 80 spores of VA-mycorrhizal fungi (Glomus, Gigaspora and Acaulospora genera) 100 g-1 soil. Almost all plants growing in desert including xerophytes do carry VAMF infections on their roots (Kiran Bala *et al.*, 1989) and the infection varied from plant to plant. In general the Young plants mostly carry mycelium with a small number of vesicles while and spores were common on mature roots.

Population of VAM fungi decline during fallow. Crops sown after long periods of fallow have poorly developed mycorrhizal systems and suffer from P-deficiency. VAMF infection on roots of neem was observed up to 250 cm depth, but the intensity of VAMF infection was inversely proportional to the availability of water. VAM-fungi also enhance nodulation and N2-ase activity in legumes (Tarafdar and Rao, 1997). Positive and a significant interaction between *Rhizobium*-VAM fungi was also observed.

Available Potassium

The arid soils are well provided with available potassium (70-890 kg ha-1). The total K content in arid region soils ranged between 980-1890 mg 100 g-1 with an average value of 1489 mg 100 g-1 soil. Major proportion of total potassium in arid soil is present as mineral form followed by interlayer, non-exchangeable and water-soluble form. These forms beside related to each other are correlated with sand and silt fraction and K resistance to depletion (Choudhari and Pareek, 1976).

Arid soils of Rajasthan contain (30 to 1270 mg kg-1) HNO3-soluble and (20 to 1120 mg kg-1) fixed potassium. Some dune and interdunal soils show negative K fixation (Aggarwal *et al.*, 1979; Joshi *et al.*, 1982a). The K fixation capacity was related with the clay content, K-saturation and weathered Kbearing minerals (Talati *et al.*, 1974; Joshi 1986a; Mathur *et al.*, 1981 and Dutta and Joshi, 1993). Kalyansundaram *et al.* (1993) reported only 5.4% soils of north Gujarat as low, 55.2% medium and 38.5% high in available potassium. In the irrigated north-west plain soils of Rajasthan Mathur *et al.* (2006) reported wide variation of potassium content within the soil group i.e. 277 to 488 kg K2O ha-1 in Torrifluvent, 215 to 389 kg K2O ha-1 in Torripsamment and 200 to 374 kg K2O ha-1 in Calciorthid soil group.

Reserve

K constituted 8 to 15% of the total K. K fixing capacity of arid zone soils are lower than of adjoining areas (Choudhari and Jain, 1979). The activity ratio for different arid soils was in the range of 3-21 x 103 mol L-1. Labile potassium, potassium adsorbed on specific and non specific sites, potassium buffering capacities and K potential were generally lower in dune inter-dune and sandy – plain soils than in the medium fine textured alluvial soils (Dutta and Joshi, 1990, 1992; Joshi 1992, 1993). In the soils of Saurastra region water soluble (0.003-0.21 m.e 100 g-1), exchangeable (0.03- 2.0 m.e 100 g-1), non exchangeable (0.32-21.7 m.e

100 g-1) and total (1.1-20.3 m.e 100 g-1) forms of potassium were observed (Patel *et al.*, 1993). Increase in CaCO3 content beyond 30% enhanced the K fixation. Different K parameters, decreased with increase in clay content (Patel *et al.*, 1993).

Secondary Nutrients (Calcium, Magnesium and Sulphur) Calcium is an essential element present in varying amount 0.1 to 5.2% in the arid soil. However, higher amount 15-20% (as CaO) is present in some lower layers of many soils. The exchangeable calcium which is indicator of availability, ranged between 1.5 to 20 m.e. 100 g-1 soil. Fine textured Aridisols of Pipar and Pali contain 15 to 20 m.eq 100 g-1 whereas dune and coarse textured contain between 1.5 to 4.0 and 2.2 to 4.0 m.eq 100 g-1 soil. Magnesium is present in various forms in arid region soil with mineral forms Magnesium is present in various forms in arid region soil with mineral forms exchangeable Fe was 2.2-5.6 ìg g-1 in different arid soils. In the arid soils of Punjab reducible and exchangeable Fe were in ferrous forms, whereas in Rajashthan and Haryana soils it was in ferric form (Shukla and Singh, 1973).

In the arid soils of Rajasthan about 40% soil samples contained 2-5 ppm and 54% samples had 5 to 10 ppm DTPA-Fe (Joshi and Dhir, 1983b). In Haryana soils of Mahendragarh district were well provided, whereas 25% in Sirsa and 51.6% in Hisar district were deficient in iron (Shukla *et al.*, 1975). Iron deficiency reported in Bhatinda (38%), Faridkot (40%) and Sangrur (11%) districts of Punjab. In Jamnagar, Kachchh, Banaskantha and Mehsana districts of Gujarat 8-23% soils were deficient and 45-65% samples marginal in available iron (Dangarwala *et al.*, 1983).

Manganese

Total manganese content in arid soils ranged between 250-875 ìg g-1. Reducible and active forms varied between 3 to 123 on 4 to 128 ìg g-1, respectively (Johri *et al.*, 1978). However, lower contents were reported for Jaisalmer and Barmer soils (Joshi and Dhir, 1982). DTPA extractable Mn in arid soils of Rajasthan varied from 1.1 to 25 ìg g-1 (Dhir *et al.*, 1983). These values were comparable with similar soils of Punjab (Arora and Sekhon, 1981) but the Haryana soils contain higher amount of Mn (6.4-73.8 ppm; Shukla *et al.*, 1975). DTPA extractable Mn in the arid soils of Gujarat showed wide range (0.8-104.4 μg g-1) and only <5% samples were deficient (Dangarwala *et al.*, 1983; Dangarwala and Patel, 1996). In arid Rajashthan 23% samples were below the critical limit, most of which were from the dune and interdune soils (Joshi and Dhir, 1983b). Available manganese was deficient in 29% samples of Hisar and marginal in 40% samples of Sirsa district.

The soils of Mahendragarh district were adequate in available manganese.

Zinc

Content of zinc in arid region soils often ranged from 41 to 130 μg g-1. Further, the dune field soils showed higher content compared to Chirai, Khajwana, Gajsinghpur and Pipar soils. Arid soils of Rajasthan are in general well provided with available zinc while deficiency could be encountered only in 13.5 and 9.6 samples in Barmer and Jailsalmer districts, respectively. Wide variations in the contents of exchangeable (0.24-1.28 μg g- 1) and DTPA soluble (0.27-2.36 μg g-1) have been reported (Joshi *et al.*, 1982b; Sharma *et al.*, 1983.

The exchangeable and DTPA soluble forms were not associated with clay, silt and organic carbon. Exchangeable and DTPA soluble forms of zinc showed irregular distribution with depth (Joshi and Dhir, 1981). Sharma and Kolarkar (1983) observed slightly higher values of DTPA Zn (0.83-1.97 μg g-1) in Jamnagar soils.

Dangarwala *et al.* (1983) reported zinc deficiency in 44, 39 and 13% samples in Banaskantha, Jamnagar and Kachchh districts, respectively. The studies revealed that about 80% samples of Hisar, Mahendragarh and Bhiwani districts of Haryana and 25% samples of similar soils in Punjab are deficient in zinc (Anonymous, 1982-83, Arora and Sekhon, 1981). The Zn adsorption by sandy soils follwed the Langmuir adsorption equation.

The quantity of Zn adsorbed by sandy soils was much less than the medium fine textured soils (Joshi *et al.*, 1983b, Joshi and Sharma, 1986). Significant relationships between the quantity, intensity and supply parameters indicated that sandy soils could maintain higher level of available Zn. However, the calcareous soils had low adsorption maxima and high bonding energy constants than the non calcareous soils. Strong affinity of Zn with CaCO3 indicated low availability of Zn in these soils (Joshi, 1996).

Copper

Contents is arid soils varies from 30 to 94 μg g-1 and medium to fine textured Pipar and Pali contained higher amounts than coarse textured Chirai and dune soils. Most of the arid soils of Rajasthan, Punjab and Haryana appeared sufficient in Cu (Dhir *et al.*, 1983; Singh, 1983). However, Takkar *et al.* (1976) reported 22 and 25% samples deficient in Cu in Sangrur and Gurudaspur district respectively. In the Banaskaantha district of Gujarat 22% deficient samples (Dangarwala *et al.*, 1983). The exchangeable (0.23-1.56 μg g-1) DTPA extractable forms of copper showed wide variations in different arid soils (0.28-1.25 µg g-1) (Joshi *et al.*, 1982b; Sharma *et al.*, 1985; Dangarwala *et al.*, 1983).

Micronutrient Transformations and their Relationship with Soil Properties

The transformation of micronutrients in soils depends upon soil pH, soil texture, type of clay mineral, organic matter, salinity/ alkalinity, quality of irrigation water, soil moisture and temperature. In arid soils, the DTPA extractable Fe was significantly and positively correlated with clay, silt and organic carbon and negatively with soil pH (Sharma *et al.*, 1983) and CaCO3 (Joshi *et al.*, 1983b). Most of the Fe deficiency in arid soil is associated with high CaCO3. Total Mn was not related with any of the soil parameters, but HCl soluble Mn was significantly related with clay, silt and organic carbon.

The reducible Mn was not governed by pH and CaCO3 as by clay and organic carbon (Joshi *et al.*, 1981). Free Mn was negatively related with pH and CaCO3 and positively with organic carbon. The mean values of reducible, exchangeable and DTPA soluble Mn decreased with increase in soil pH, CaCO3, organic matter (Joshi and Dhir, 1983a) and finer fraction of soils (Joshi and Dhir, 1983b, Lal and Biswas, 1973). However, for Zn, only HCl soluble fraction was associated with clay, silt and organic carbon, but exchangeable and DTPA soluble forms were not related with any of the soil parameters (Joshi *et al.*, 1982b; Sharma *et al.*, 1983). The HCl soluble Cu was significantly related with clay, silt, organic carbon and CaCO3 content of soils. Available Cu was significantly related positively with clay, silt and organic carbon and negatively with pH. Exchangeable Cu was critical in arid brown and marginal in sierozem soils of Punjab and Haryana, respectively (Grewal *et al.*, 1969; Singh and Shukla, 1984; Singh *et al.*, 1988). Medium textured soils contained slightly higher amounts of DTPA soluble Cu than dune, interdune and sandy plain soils.

There was marked effect of saline/high RSC irrigation water on availability of micronutrients in soils. Soils irrigated with saline and high RSC waters were low in available micronutrients (Joshi, 1988, 1990b). The contents of HCl soluble, free and reducible forms of Mn in salt affected soils were comparable with the associated non saline soils, but their predictability was much reduced (Joshi *et al.*, 1988a). However, little difference was observed in HCl-Fe and free Fe in salt affected and associated non saline soils. Due to the effects of different soil parameters in salt affected soils, the HCl-Fe was better predictable but free Fe had low predictability (Joshi *et al.*, 1988a). Exchangeable

Fe in salt affected and associated non saline soils were comparable. The salt affected and associated non saline soils did not differ in various forms of Zn, but high salinity appears to vitiate their predictability (Joshi *et al.*, 1988b). Though there was no difference in the contents of HCl soluble and exchangeable Cu in salt affected soils and the associated nonsaline soils, but their predictability was considerably low (Joshi *et al.*,1988b). Adsorption/desorption behaviour of micronutrients in arid soils.

The adsorption of Zn and Cu in arid soils followed Langmuir adsorption equation. Because of the low clay/organic matter, low cation exchange capacity, high base saturation and calcareous nature of sandy soils, they have characteristic adsorption and release behaviour for these nutrients. The quantity of Zn adsorbed by sandy soils was much less than medium/fine textured soils (Joshi *et al.*, 1983b; Joshi and Sharma, 1986).

Significant relationship between quantity, intensity and supply parameters indicated that sandy soils could maintain higher level of available Zn. Strong affinity of Zn with CaCO3 indicated low availability of this nutrient in such soils (Joshi, 1996). However, Cu could be adsorbed at low concentration but with increasing addition, there was precipitation of Cu rather than multilayer adsorption (Joshi, 1995). The adsorption of Cu was less in fine sandy than loamy soils (Joshi, 1986b).

Micronutrients in Plants and their Recycling in Arid Soils

It is possible to estimate the extent to which the micronutrient cations are recycled through the decomposition of plant materials from assumptions about the dead materials and the concentration of cations. Rainfed crops and grasses/shrubs of arid region have higher amount of micronutrient content as compared to irrigated crops. Among grasses, substantial higher amount of copper was observed in bharut (*Cenchrus biflorus*) and sewan (*Lasiurus sindicus*) grass. Zn was found deficient in irrigated crops (groundnut, wheat, and mustard), but was not the case with rainfed crops (Soni *et al.*, 2003).

Among the perennial components, the return of Fe is maximum through trees (Total Fe=381 ppm) followed by shrubs (Total Fe=251 ppm) and grasses (Total Fe=235 ppm) of arid region (Sharma *et al.*, 1985). As regards Mn, arid zone grasses have highest concentration of Mn followed by tree species. Grasses have also significantly higher values of Zn and Cu as compared to trees and grasses (Sharma *et al.*, 1984). Fleming (1973) and Dhir *et al.* (1985) also stated that arid zone grasses are significantly low in Mn and quite high in Zn and Cu.

Various soil degradation processes viz., Aeolian hazards, water erosion including floods have detrimental effect on micronutrient availability (Raina and Joshi, 1994).The soils under natural pasture were low in available micronutrients than the soils under sown ungrazed pastures (Raina and Joshi, 1991). Comparatively higher amount of micronutrient was observed in natural rangelands followed by arable rain fed cropping and minimum in irrigated conditions (Soni *et al.*, 2006).

Considering the status of natural rangelands as ideal, the percent losses of Fe, Zn, Cu and Mn showed that reduction in available micronutrients was more in irrigated lands as compared to rain fed cultivated lands. On an average, Fe, Zn, Mn and Cu were reduced by 32.4, 24.0, 23.8 and 12.0% in irrigated and 9.2, 8.4, 10.4 and 6.8% in rain fed cultivated soils over natural rangeland systems, respectively (Soni *et al.*, 2006). Build up of all the four micronutrients (Mn, Zn, Cu and Fe) was more in medium and fine textured soils under *Oran* system as compared to cultivated lands.

Nutrient Management in Arid Soils

Nutrient use efficiency is evaluated using metrics that reflect crop uptake of fertilizer added in the current growing season. The problem of low nutrient use efficiency is conventionally viewed as a consequence of temporal asynchrony and spatial separation between applied nutrients and the crop. As a result, efforts to improve nutrient utilization efficiencies have emphasized on improved delivery of nutrients to the root zone during the period of crop uptake through modifications such as banding, fertigation, split fertilizer applications, etc. To increase crop access to N fertilizer, a variety of additives have been developed that inhibit urease activity, nitrification and denitrification. These approaches have been extremely successful in terms of increasing nutrient use efficiency and maximizing yields; however, their utilization on mass scale has met with limited success.

Despite more than 30 years of concentrated effort, mass balances indicate that annual N and P inputs consistently exceed harvested exports by 40 to >100% resulting in substantial losses of these nutrients to the environment. Thus, it is evident that global challenge of meeting increased food demand and protecting environmental quality will be won or lost in cropping systems (Cassman *et al.*, 2002). Major limitation of most of the approaches tried in past three decades is lack of their applicability on large scale on small farm holdings. Only two approaches hold promise for application on mass scale to increase nutrient use efficiency (1) application of optimum quantity and (2) internal nutrients circulation.

Application of Optimum Quantity

One of the first step necessary for achieving higher fertilizer use efficiency (FUE) is to apply the right amount of fertilizer. But the long term estimations of N requirement of crops under rain-fed conditions may be very difficult as the yield levels vary from zero to three times of average yield. Using average yields in arid regions is too conservative and may actually result in lowering average yields with time because of insufficient nutrient availability for the very favourable years and on contrary, use of relatively high yield goals results in excess N applications in most years and can greatly reduce profit.

The current concern over the potential for excess N to degrade the environment also makes this alternative unacceptable. Tucker (1988) presented three ground rules to arrive at logical yield goals. These include choosing yield goals based upon: (1) highest yield within the past 5 years with good crop management (2) yield goal set a 1.5 times of long term average, and (3) yield goal based on soil capabilities as defined in Standard Soil Surveys, using yields of highest yield obtaining farmer's in the vicinity on the same kind of soil.

Uncertainties in soil nitrogen (N) supply in rainfed agriculture and crop N demand present a challenge to scientists in deciding on N fertilizer rates. Number of field studies have documented that the improvements in N use efficiency is possible with site-specific N management approaches. Lobell (2007) has presented a general model of N rate decision-making which computes the optimal N rate that maximizes expected profit given uncertainties in N supply and demand.

The cost of uncertainty is measured as the difference in N rate when soil N supply and crop N demand are unknown versus known perfectly. Eliminating uncertainty in soil N supply (but not crop demand) reduces average N rates by 5–40% in different crops while perfect knowledge of potential crop N demand (but not soil supply) reduces rates by 3-10%. Simultaneous knowledge of both factors reduced N rates by significantly more than the sum of their individual effects. This approach is very different from most extension services in India which provide a single, standard fertilizer recommendation for an entire district or region. Farmers apparently have few guidelines for adjusting N-fertilizer amount to account for the large differences in the indigenous N supply, indicating the need for a 'field-specific' approach to N management. Fieldspecific N management could lead to substantial reductions of N rates without yield loss in a wide range of cropping systems, thereby improving profitability and environmental quality.

Internal Nutrient Circulation

The approach comprises of various conventional practices like recycling of crop residues, either through direct application or as manure and cultivation of legumes often referred as integrated nutrient management in past. The practice of effective use of inorganic and organic sources of nutrients together in a proper proportion not only reduces the requirement of inorganic fertilizers, but also improves physical conditions of soil, enhances water retention capacity and its availability in the soil. Apart from this, the biological properties of soil are also improved considerably (Sharma *et al.*, 2005, 2008, 2009) and fertilizer use efficiency is also increased.

The results of several studies have shown that the combined use of fertilizer and farm yard manure (FYM) is very much essential in rainfed areas for sustaining moderate to high yields and achieving higher nutrient use efficiencies. Besides, application of FYM and composted organic wastes and other organics improve yield stability in rainfed areas (Aggarwal and Venkateswarlu, 1989; Venkateswarlu and Hegde, 1992; Singh *et al.*, 2000). Singh *et al.* (1981) observed that under arid conditions of Jodhpur continuous application of sheep manure in general gave substantially higher yields than the application urea alone. Rao and Singh (1993) showed that substitution of 50% of fertilizer requirement by FYM resulted in yield levels nearly similar to those obtained with complete fertilization.

Aggarwal and Praveen-Kumar (1996) on the basis of a seven year study on arid soils showed not only a beneficial effect of FYM application but also a synergistic effect of simultaneous application of FYM and inorganic fertilizers on crop yield. FYM application increases the utilization efficiency of fertilizer N, however, improvement in soil fertility after FYM application is a very slow process. Nutrient content of pearl millet, sorghum stover is relatively low, stover can contribute to the productivity of the soil. Such residues must be managed carefully, however, because N can be immobilized at the time of peak maize N requirements, resulting in poor crop growth. In rainfed areas, cereal stover is often fed to livestock, and manure is applied in field. This way of recycling the residues is more beneficial for crop than their direct application in field. Losses of N from such systems are often high. Cattle manure is applied in a dried, aerobically decomposed form, often with a high sand content and N content that is frequently around 1% or less.

Research shows that the most efficient use of manure is to combine it with some inorganic fertilizer. Ladd and Amato (1985) and Snapp and Silim (2002) showed that the most promising route to improving inorganic fertilizer efficiency in cropping systems is by adding small amounts of high-quality organic matter (possessing a narrow C/N ratio and a low percentage of lignin) to soils. It provides readily available N, energy (carbon), and nutrients to the soil ecosystem, and improve structure, and increase soil microbial activity and nutrient cycling and reduce nutrient losses from leaching and denitrification (De-Ruiter *et al.*, 1998). Soil microbes are valuable not only because they supply nutrients directly, but because they enhance the synchrony of plant nutrient demand with soil supply by reducing large pools of free nutrients (and consequent nutrient losses from the system). Thus, microbes maintain a buffered, actively cycled nutrient supply.

Incorporation of crop residues and natural vegetation in soil improve microbial activity during decomposition. Also adhesive action of decomposed products improves soil aggregation, hydraulic conductivity and moisture retention (Venkateswarlu, 1984, 1987, Gupta, 1984; Gupta and Gupta, 1986). Leaving the crop residues in soil generally have a positive effect on grain yield (Aggarwal *et al.*, 1996). Crop residues are also efficient source of nutrient like other organics viz., cattle manure and compost. However, Aggarwal *et al.* (1996) did not find any significant change in the yield of succeeding crop of pearl millet after addition of crop residues with wide C:N ratio, whereas it was significant after incorporation of residues with narrow C:N ratio.

Legume rotations are an important practice for restoring soil fertility on larger land holdings. The amount of N returned from legume rotations depends on whether the legume is harvested for seed, used for forage, or incorporated as a green manure. Estimated net N return of 23-110 kg ha-1 from pigeon pea, 23-50 kg ha-1 from dolichos beans, and 25-60 kg ha-1 from groundnuts has been reported by Giri and De (1980). Mishra (1971), Mann and Singh (1977), Singh (1980) and Singh *et al.* (1985) also reported that in arid soils pearl millet-cluster bean rotation gave higher yield than of continuous cultivation of pearl millet due to improved soil fertility (Praveen-Kumar *et al.*, 1996). Singh *et al.* (1985) in a long-term study found an increase soil organic carbon by 12% and available soil P by 25% after legume cultivation. Singh and Singh (1977) on the basis of a longterm study reported that cultivation of green-gram in rotation with pearl millet supplied with 20 kg N ha-1 gave similar yield as with application of 40 kg fertilizer N ha-1. Singh *et al.* (1985) observed that rotation of pearl millet with green gram or clusterbean was better than its rotation with moth bean.

Praveen-Kumar *et al.* (1996) reported that higher yield of pearl millet when it was preceded by clusterbean cultivation than green gram.

Beneficial effect of legumes to pearl millet also depends on the number of seasons of their cultivation prior to pearl millet. The intercropping of pearl millet and legumes in arid soils has also shown promising results (Mishra, 1971; Singh and Joshi, 1980; Singh *et al.*, 1978).

Major quantity of N i.e, upto 70% of total N uptake, can be taken up by the pearl millet within first 30 days of growth. Though it slows down thereafter, but continues gradually till the grain filling stage. This high N demand cannot be met from the soil and thus pearl millet respond favorably to the N addition (Aggarwal and Vekateswarlu, 1989), which ranged from 7.0 to 18.0 kg grain kg-1 N with higher values in good rainfall years. Singh *et al.* (1981) have reported that the yield of pearl millet doubled with the application of 40 kg N ha-1. Similar results have also been reported by Singh *et al.* (1981). Aggarwal and Praveen-Kumar (1996) on the basis of a seven year long study reported significant response to application 80 kg N ha-1 only in the years of good rainfall. Comparison of different N fertilizers showed that, maximum yields were recorded with ammonium sulphate. But due to high cost of ammonium sulphate, urea has become the major source of N in arid regions even though utilization efficiency of N from urea by pearl millet is very low. Various studies in CAZRI Jodhpur have revealed that the mixing of elemental S with urea (Aggarwal *et al.*, 1987) or application of small quantity of ammonium sulphate before the application of urea (Praveen- Kumar and Aggarwal, 1988) increases its efficiency. Praveen-Kumar and Aggarwal (unpublished data) also showed that NH4 : NO3 ratio in a N fertilizer also effect its utilization efficiency and observed maximum yield levels and highest utilization efficiency from the applied N fertilizer when NH4 : NO3 ratio in the fertilizers was 3:1. Besides ammonia volatilization denitrification is now being considered an important pathway of N loss (Peterjohn, 1991). Due to different loss mechanisms major part of N applied, if not used by crop is lost from soil. As a result residual effect of fertilizer N is generally not observed (Aggawal and Praveen-Kumar, 1996). Praveen-Kumar and Aggarwal (Unpublished data) demonstrated this in a field experiment wherein 20, 40, 60 and 80 kg N ha-1 was applied in different plots at the time of sowing and the crop was harvested after 20, 40 and 60 days of growth symbolizing complete crop mortality at early, medium and late stage of drought, respectively. Next year pearl millet was again grown in these plots without addition of N fertilizer to study the residual effect. The residual effect was compared with the yield in some other plots where fertilizer @ 0, 20 and 40 kg N ha-1 was added. Results indicated comparable grain yield in most plots where residual effect was studied to those in control plots where no fertilizer was added. This suggested that fertilizer N applied previous year did not show significant residual at any level of fertilizer application and stage of crop harvest. Since droughts occur frequently in arid regions because of which crop often fails, therefore, farmers consider application of N fertilizers a risky input and hence the need for its better management. Split application of N fertilizers is therefore, considered a risk avoiding strategy which results in the higher yield as compared to their one time application, increased utilization efficiency and lower risk. Praveen-Kumar and Aggarwal (1997, 1998) observed basal application of fertilizer N may be avoided if clusterbean has preceded the pearl millet.

Biodiversity in Soils

Soil biodiversity is an important resource that provides ecosystem processes essential to the functioning of natural and global systems. Biota occupy remarkably different habitats within profile of arid soils with some species living on decaying biological matter, some in the rich organic matter horizon, and some in deeper layers and the species composition may vary even at millimeter scale depending on type and amount of vegetation and soil texture. Soil organisms moderate numerous ecosystem processes (e.g., decomposition, C and N transformation, hydrologic cycles). Changes in landuse, climate and atmospheric can alter species composition within the soil community, and may impact ecosystem function. Current priorities in the study of soil biodiversity include:

(1) understanding the functions of rarely sampled fauna so that species estimates and global distributions are based on an accurate and current database;

(2) determining which habitats are most vulnerable for soil biodiversity loss, e.g., where the 'hot spots' of biodiversity are, and which habitats and what time frames are most amenable to restoration;

(3) synthesizing data and determining which invertebrate and microbial 'species' are key to ecosystem processes;

(4) identifying gaps in knowledge of multispecies interactions and their influence on ecosystem functioning;

(5) collaborating on long-term (more than 3 years) experiments to examine effects of global changes on aboveground-below-ground biodiversity linkages and ecosystem functioning; and

(6) determining, based on natural history information, which species are more likely to be invasive if introduced, and using this potentially to reduce spread and to identify threats to other species.

Macro-fauna

The arid soils are home to diverse life forms adapted to hot and dry milieu. Soil organisms contribute in formation of organic matter and nutrient cycling. The burrowing invertebrates create macropores that allow rapid flow of water into or through soil. The tiny arthropods produce fecal pellets that are mixtures of soil and organic matter. These become stable soil aggregates. The arid soil meso and macro fauna are constituted largely by insects, mites, earthworms and millipedes. Predators in the soil food web include scorpions, centipedes, spiders, mites, some ants and beetles. They check the population of many soil biota. The reptiles and rodents represent the soil dwelling vertebrates, the former play a predatory role while the latter thrive on a variety of food resources.

Soil compaction, low vegetation or plant litter reduces the number of soil arthropods. Their level of activity depends on micro-environmental conditions including temperature, moisture, aeration, pH, pore size, and types of food sources. Earthworms move soil from lower strata up to the surface and move organic matter from the soil surface to lower layers. Suthar (2002) studied the earthworm resources of arid Rajasthan and their utilization in vermitechnology. Suthar (2009) recorded occurrence of nine species of earthworms from Jodhpur district. These included *Pontoscolex corethrurus,* Amynthas morrisi, Metaphire posthuma, Lampito mauritii, Perionyx sansibaricus, Ocnerodrilus occidentalis, Dichogaster bolaui, Ramiella *bishambari* and *Octochaetana paliensis.* Diversity and abundances of earthworms in conventional, integrated and organic agro-ecosystems was also measured. Maximum numbers of earthworm were found in integrated farming system and abundance of earthworms in arable system was directly related to the management practices (Suthar, 2009). There have been but few studies on the soil macro faunal associations in different production systems of the arid regions and their contribution to soil fertility. Tripathi *et al.* (2005) reported that Acari, Myriapoda, Coleoptera, Isoptera and Collembola arthropods as the major soil faunal groups in silvipastoral systems of arid zone. Relative densities of Acari, Myriapoda and other arthropods were highest in silvopastoral systems and those of Coleoptera, Isoptera and Collembola were highest in pure grass plots. Termites are another important arthropods in the soil.

Termites aid in crop residue breakdown, enhance the porosity of soil, increase the soil's infiltration capacity and affect nutrient cycling by concentrating organic matter in the nests. However, their beneficial role is shadowed by the damage they cause to the cultivated plants. All the known species in the arid regions are subterranean. This group of arthropods has been studied fairly well owing to their economic importance in all fields. Nearly two scores of termite species have been recorded from the arid region (Rathore, 1996). Parihar (1981) collated the available information about termites and their management.

The known termite fauna of the arid parts of Rajasthan, Gujarat, Haryana and Punjab are represented by the following species: *Amitermes belli, Anacanthotermes macrocephalus, Angulitermes* jodhpurensis, Coptotermes heimi, C. kishori, Eremotermes fletcheri, E. neoparadoxalis, E. paradoxalis, Heterotermes indicola, Incisitermes didwanaensis, Microcerotermes cameroni, M. heimi, M. laxmi, M. palestinensis, M. raja, M. sakesarensis, M. tenuignathus, Microtermes mycophagus, M. obesi, M. unicolor, Odontotermes assmuthi, O. bellahunisensis, O. brunneus, O. dehraduni, O. distans, O. feae, O. giriensis, O. girnarensis, O. guptai, O. gurdaspurensis, O. latiguloides, O. microdentatus, O. obesus, O. paralatiguloides, O. redemanni, O. sasangirensis, O. O. wallonensis, O. lokanandi, O. indicus, Psammotermes rajasthanicus, Speculitermes cyclops, S. sinhalensis, Synhamitermes *quadriceps, Trinervitermes biformis* and *T. fletcheri.*

Ants also modify soil properties through their nesting and foraging activities. While most of the ants found in arid regions are foragers, a few are predatory and some tend sucking insects. Still there have not been extensive studies about the foraging by ants in the arid regions. The available information is mostly restricted to documentation of the occurrence of ants in the arid tracts (Tak and Rathore, 1996; Tak, 2008), which include *Acantholepis frauenfeldi, Anochetus punctiventris, A. sedilloti, Camponotus* angusticollis, C. compressus, C. irritans, C. mitis, C. taylori, Cataglyphis *bicolor setipes, Crematogaster brunnea contemta, Dorylus labiatus, D. orientalis, Meranoplus bicolor, Messor barbarus humalayana, Monomorium* atomus, M. criniceps, M. destructor, M. glaber, M. indicum, M. pharaonis, M. scabriceps, M. wroughtoni, M. gracillimum, M. latinode, Pheidole sulcaticeps, *P. wroughtoni, Sima rufonigra* and *Tapinoma melanocephalum.*

Coleoptera is another important order of insects associated with soil. The grubs of most beetles are inhabitants of soil, plant roots or decaying matter. Some of these are also predators of injurious insects. Being of agricultural importance, beetles and weevils have received attention of researchers. However, there is no account of their benign contribution in enriching the soil. Pal (1977) had given an account of the white grubs prevalent in the arid regions. Tripathi *et al.* (2008) studied the diversity and abundance of beetles in some silvipastoral systems. The more common Scarabaeids in the arid region are species of the genera *Lachnosterna* (*Holotrichia*), *Anomala*, *Adoretus*, *Rhinyptia*, *Schizonycha*, and *Onthophagus*.

Bembidion luniferum, *Asaphidion triste*, *Leitus indus* and *Scaritus* species are more common Carabids. There is a large number of other beetles whose grubs inhabit the arid soil. The grubs and adults of dung beetles contribute profusely in decomposing the organic matter in arid soils. The beetles process large amounts of animal dung into nodules or balls, and roll these into subterraneous chambers or tunnels where they are degraded, thereby increasing soil fertility. They also prevent the use of decaying materials by other pest insects. Eighty-five species of these beetles have been documented from the arid regions belonging to *Onthophagus, Caccobius, Onitis, Copris,* Scarabaeus, Catharsius, Gymnopleurus, Oniticellus, Phallops, Drepanocerus, *Heliocopris* and *Chironitis* genera (Ram-Sewak, 2009).

The Collembolan insects contribute significantly in litter decomposition. The majority of springtails feed on fungal hyphae or decaying plant material. But not enough information is available about them from the Indian arid soils. Seven species of Collembola including *Xenylla obscura* and *Isotomodes dagamae* were recorded from Sriganganagar district by Faisal and Ahmad (2005). Their existence and establishment has been attributed to the change in land use pattern due to the development of canal irrigation. Although a large number of species of soil mites exist in the arid regions, there have not been any systematic attempts to study these organisms.

Sanyal (1996) while giving an account of the soil mite fauna from Rajasthan has listed only three species *viz.*; *Eugamasus* sp, *Galumna flebellifera* and *Microsmaris pinnata* collected from the drier tracts. There have been many studies on rodents in the arid regions, their damage and control strategies, but no precise estimates are available about their contribution to soil fertility, if at all. The soil inhabiting rodents in the arid zones include *Bandicoota bengalensis, Gerbillus gleadowi, G. nanus,* Meriones hurrianae, Mus booduga, M. cervicolor, M. musculus, M. platythrix, *Nesokia indica, Rattus gleadowi, R. meltada pallidior* and *Tatera indica*. There is a need to generate more information about the soil meso and macro fauna that influence soil fertility under arid conditions and the ways to enhance their activities.

Soil Microorganisms

Microbial activity in desert soil is highly dependent on characteristics such as temperature, moisture and the availability of organic carbon (Parkar *et al.*, 1984; Rao and Tarafdar, 1998). Of these, moisture availability is the major factor affecting microbial diversity, community structure and activity. Populations of aerobic bacteria in deserts across the world are reported to vary from <10 in Atacama desert to 1.6´107 g-1 in soils of Nevada (Skujnis, 1984). Populations in Indian arid zone reported to have relatively smaller (1.5´102-5´104 g-1 soil) (Venkateswarulu and Rao, 1981; Kathuria, 1998). Gram-positive spore formers are dominant and the populations do not decline significantly even during summers (Rao and Venkateswarlu, 1983).

Actinomycetes may constitute ~ 50% of the total microbial bacterial population in desert soils. Dominant microflora of desert soils is made up of coryneforms, i.e. *Archangium, Cystobacter, Myxococcus, Polyangium, Sorangium* and *Stigmatella*; sub-dominant forms comprise *Acinetobacter, Bacillus, Micrococcus, Proteus* and *Pseudomonas*. Cynobacteria also contribute significantly (0.02 to 2.63 ´ 104 g-1 soil) to the biota of hot arid regions in terms of primary productivity and nitrogen fixation (Bhatnagar *et al.*, 2003). The dominant cynobacterial forms of Thar desert are *Chroococcus minutus, Oscillatoria pseudogeminata* and *Phormidium tenue*; *Nostoc* sp. dominates amongst heterocystous forms. Fungal populations as viable propagules range from 0.5 to 14.7´103. The dominant genera include *Aspergillus, Curvularia, Fusarium, Mucor, Penicillium, Paecilomyces, Phoma* and *Stemphylium*. Xeric mushrooms such as *Coprinus, Fomes, Terfezia* and *Teramania* and arbuscular mycorrhizal fungi such as genera *Glomus, Gigaspora* and *Sclerocystis* have also been reported from desert (Trappe 1981; Pande and Tarafdar, 2004). Rhizosphere is an important site of microbial activity in desert soils, since it provides ample carbon substrate in an otherwise organic matter poor arid soil. Generally R:S (Rhizosphere : soil) ratio is high in arid soils for nearly all metabolic types of bacteria (viz. heterotrophs, dizotrophs, cellulites and nitrifiers) and fungi in most plants studied (Kathuria, 1998; Singh and Tarafdar, 2002). The soil zone penetrated by fine roots and held together by mucilage is called the rhizosheath.

Bacillus polymyxa and the fungus *Olpidium* are found associated with rhizosheaths. *Ancalomicrobium* and *Hypho-microbium*- like organisms are also present (Wullstein and Pratt, 1981). Rhizosheaths are important because of the associated diazotrophs and enhanced water retention and nutrient uptake (Watt *et al.*, 1994). In the rhizosphere of desert plants mycorrhiza play a very significant role in plant nutrition and stabilization (Tarafdar and Rao, 1997, 2002; Tarafdar and Gharu, 2006). Most species in families of *Asteraceae, Fabaceae, Poaceae, Rosaceae* and *Solanaceae* usually form endomycorrhizal associations in arid habitats (Skujins, 1984). Trappe (1981) listed 264 plant species from arid and semi-arid environments that had mycorrhizal-based root colonization and approximately 25% species exhibited specific associations with endomycorrhizal fungi. *Glomus deserticola* is indigenous to many desert soils (Bhatnagar *et al.*, 2003). Kiran Bala *et al.* (1989) reported >50% infection by vesicular-arbuscular mycorrhizal (VAM) fungi in 17 tree species of the Indian desert, with genera *Glomus* and *Gigaspora* being dominant.

Opunitia and *Euphorbia* showed considerable root infection. In desert, incidence of arbuscular mycorrhiza (AM) infection varies with the availability of water (Tarafdar, 1995) and with composition of plant community (Kiran Bala *et al.*, 1989). Mycorrhiza also help in desert reclamation and soil stabilization. They link soil particles to each other and to the roots in part by producing glomalin, an important glue that holds aggregates together. Panwar and Vyas (2002) reported beneficial effects of *Acaulospora mellea, Gigaspora margarita, G. gigantean, Glomus deserticola, G. fasciculatum, Sclerocystis rubiformis, Scutellospora calospora* and *S. nigra* on *Moringa concanensis* and proposed use of such AM fungi in conservation of this endangered multipurpose tree species in Indian desert. Crusts on soils are formed by microorganisms and microphytes. These are variously called as cryptogannic/cryptobiotic/biological/cynobacterial/ microphytic crusts (West, 1990). A consortium of green algae, cyanobacteria, lichen, fungi, bacteria, diatoms, mosses and liverworts form cryptobiotic crusts. Mosses and liverworts seem to favour the comparatively more mesic sites. Cynobacteria favours the harshest sites, and lichens dominate in intermediate sites.

Cynobacterial crusts generally dominate poor sandy soils. Lichens increase proportionately with carbonates, gypsum and silt content of the substrate (Gracia-Pichel *et al.*, 2001; Budel, 2002). *Microcoleus chthonoplastes* is dominant in saline sand crusts (Gracia-Pichel *et al.*, 2001). The most common genera are *Microcleus* (*M. Chthonoplastes, M. paludosus, M. sociatus* and *M. vaginatus*) and *Nostoc* sp. Other forms are *Calothrix, Lyngbya, Oscillatoria, Phormidium, Scytonema* and *Tolypothrix*. Common green algae of soil crusts are *Chlorella, Chlorococcum, Coccomyxa* and *Klebsormidium*. Dominant fungal genera in crusts are *Alternaria, Fusarium* and *Phialomyces* whereas in non-crusted soils, *Alternaria* and *Penicillium* dominate followed by *Fusarium*. Approximately 90% of the aerobic bacterial population of cryptobiotic crusts consists of corneform bacteria (Skujins, 1984).

Diversity status of cryptobiotic crusts at Thar desert in India showed 43 morphotypes of diazotrophs in BG11-N enrichment and 71 of algae and cyanobacteria in the same medium supplemented with nitrogen (Bhatnagar *et al.*, 2003). Most frequent form was *Phormidium tenue*. In the case of diazotrophs the most frequent forms were *Nostoc punctiforme*, *Nostoc commune* and *Nostoc polludosum*. Cynobacteria presence was influenced by their plant partners. *Alternaria* sp. was the dominant fungus in these crusts. Microorganisms in soil are poorly characterized. It has been established that the genetic diversity of soil has an estimated 6400 to 38000 prokaryotic species per gram. On the other hand, less than 0.3% of the microorganisms present are culturable by standard techniques (Pettit, 2004). Only a handful of studies have looked at microorganisms in tropical soils. Microbiological properties of these soils also varied depending on the type of land use patterns. Grasslands, in general, supported higher number of microorganisms than tree plantations, cultivated fields or barren land. Stabilization of shifting and sand dunes introduction of vegetation has markedly increased the soil microflora. In general, the low organic matter content and poor moisture availability of desert soils were the major factors limiting optimum microbial activity.

Fungi like *Fusarium*, *Gliocladiu*, *Penicillium* and *Trichoderma* are stress tolerant. Majority of fungi are mesophiles with maximum growth between 25 and 30°C (*Mucor mucedo*, *Mortierella*, *Penicillium chrysogenium*), however, *Cylindrocarpon* sp. *Candida scotti* are cold tolerant (psychrotolerant) and can grow near 0°C; others are thermotolerant and grow above 40°C (*Rhizomucor*, *Thermomyces*, *Talaromyces*). Xerotolerant fungi can grow on dry material (*Aspergillus*, *Pencillium*) with low atric potential (aw) while osmotolerants grow at very low osmotic potential (*Pichia* sp.). Dung of herbivorous mammals harbors a large number of fungi, termed coprophiles, of while *Pilobolous*, *Ascobolus* and *Basidiobolus* are famous for their special shot-gun dispersal mechanism ! An interesting ecological group of fungi captures and grows parasitically on nematodes, their cysts and eggs (nematophagous); over 150 species are reported (Gray, 1998) within the genera *Catenaria*, *Dactyela*, *Harposporium*, *Monacrosporium*, *Nematophthora*, *Rhopalomyces* and *Stylopage*. Such fungi are good trapping agents in the biological control of nematodes.

Potentials of the Arid Zone Soils

Soils in the drier regions have low reserves of soil organic matter. Singh *et al.* (2007) reported 2.13 Pg carbon stock in the 0-100 cm depth of which 1.23 Pg as soil organic carbon and 0.90 Pg as soil inorganic carbon, in the arid and semi arid regions of Rajasthan. They had studied the same under different soil types. Per year 4200- 4600 kg km-2 soil organic carbon can be sequestered in the untilled soils of arid Rajasthan, if they are put under canopy cover round the year. The desert and saline soils in the Kutch peninsula is reported to have a current total carbon stock of 0.8 Pg with an estimated potentials of 1.3 Pg carbon. Uncertainties persist in the estimation of net flux of CO_2 from the soils of tropical areas due to inconsistency in land use/ land cover pattern (Chapin and Ruess, 2001; Jones *et al.*, 1998). The studies conducted at RRS, Kukma indicated soil organic carbon stock of 35.13 t ha-1 under the silvipastoral system involving *Neem* and *Cenchrus ciliaris,* 31.9 t ha-1 under Acacia and *Cenchrus ciliaris,* in the 100 cm soil depth. The carbon sequestration in soil is strongly affected by the root production (Matamala *et al.*, 2003). The trees in the arid zones put forth a large volume of below ground biomass in the form of roots. The tree roots play an important role in adding organic matter to the soil. About 25% of the total living biomass of trees is in roots which continuously add organic matter to soil by death and decay of roots. It is more important in arid zone where climax tree species like *Prosopis cineraria* that have very deep root system which can reach upto 70 m depth. Thus in arid zones large volume of carbon gets sequestered in lower layers, that have high resilience. Narain (2008) reported planting trees and grasses in the degraded lands of arid zone can help to increase the soil carbon stock from 24.3 Pg to 34.9 Pg. The studies conducted at RRS, Kukma indicated a total below ground carbon stock of 1.61 t ha-1 that constituted 23.4% of total carbon stock under silvipastural system involving *Acacia tortilis* and *Cenchrus setegerus* and 27.6% contribution by below ground portions to the total carbon stock under *Acacia tortilis* and *Cenchrus ciliaris.* The intentional controlled burning of biomass can produce charcoal or black carbon that can contribute to carbon sequestration (Goldberg, 1985). The activated charcoal, apart from locking carbon in the soil for centuries, can attribute positively to the cation exchange capacity, physical, chemical and biological properties of soil.

Chapter 5

Adaptations of Plants to Arid Environments

Environmental stresses of low and unpredictable precipitation, low relative humidity with desiccating winds, and high summer temperatures characterize climates of deserts and, coupled with low nutrient availability, produce severe limitations of plant growth. Despite such stresses, desert scrub communities often contain surprisingly large amounts of plant biomass, and possess remarkable diversity of plant growth forms.

The life form of a plant – whether annual, perennial, herbaceous, woody, or succulent – and the characteristics or its roots, stems, and leaves are presumed to be adaptations to the special conditions within a desert. Diversity of life forms may be considerable (Sonoran Desert) or very low (Atacama Desert).

Life forms may be classified into 4 major categories that represent strategies of adaptation, according to Solbrig and Orians, 1977.

Strategies

1. Drought-escaping plants – annuals which germinate and grow only when there is sufficient moisture available to complete their life cycle. Only their seeds persist during times of drought. Annuals.
2. Drought-evading plants – non-succulent perennials which restrict their growth activity to periods when moisture is available. Typically, they are drought-deciduous shrubs which go dormant or die back during dry periods.

3. Drought-enduring plants – evergreen shrubs. Extensive root systems coupled with various morphological and physiological adaptations of their aerial parts enable these hardy xerophytes to maintain growth even in times of extreme water stress. Creosote bush (*Larrea tridentata*)
4. Drought-resisting plants – succulent perennials. The water stored in their swollen leaves and stems is usually used very sparingly. Cacti

The major adaptation of both drought-escaping and drought-evading plants is an ability of accurately predict the wet season and to restrict their major growth and reproductive activities to the wet part of the year.

Drought-escaping Plants: Annuals

1. The proportion of annual species in desert floras is inversely related to the amount and reliability of precipitation in a region.
2. Life cycles of these small, shallow-rooted plants commence when there is water available.
   a. Went noted that no seedlings of any species germinated in the Sonoran Desert following a 10 mm rainfall; extensive germination occurred only after a rainfall of 25 mm.
   b. This was attributed to removal of inhibitors from the seed coats.
3. Winter and summer annuals are distinguished on the basis of optimum temperatures for seed germination. This varies from 15-18 C for winter annuals, and 25-30 C for summer annuals. These refer to temperature conditions prevailing at the time of precipitation.
4. Continued survival of annuals requires that adequate seed reserves be maintained through dry periods until conditions are once again suitable for germination.
5. Few seeds rot in the desert environment, but many are lost to seed predators, particularly rodents. Kangaroo rats (*Dipodomys merriami*) consume as much as 95% of the seed of Erodium cicutarium, an annual of the Mojave Desert.
   a. The highest seed densities are associated with wind shadows where sees are protected from further movement by the wind; lowest values were found in open sites and dry washes.
6. Seeds of most desert annuals have temperature or moisture controlled dormancy which may prevent germination, but seed

viability is initially high. Seeds may remain viable for up to 10 years under artificial conditions.

7. Mechanisms which prevent seeds from germinating all at once can increase the chance of survival of annual species. Seed heteroblasty, in which germination requirements differ for seeds produced by the same plant, has been described for some annuals in the Saharo-Arabian Deserts.
8. Since they are not restricted by water – they grow only when water is relatively abundant – they can exploit the favourable light and temperature conditions of the desert. They require none of the morphological adaptations that other strategists require.
9. Annuals' "goal" is to grow fast (large leaves, maximal photosynthesis, heavy transpiration), flower, set seed, disperse seed, and persist through the dry period as a seed.

Drought-evading Plants

1. Drought-deciduous plants; they drop their leaves, and sometimes their stems during periods of drought.
2. Leaf production from mid-Feb. to April, and a leaf canopy persists until drought conditions become extreme in summer.
3. Leaves may be polymorphic.
   a. Leaves developed in early spring when water is available and temperatures moderate, are large and green. High photosynthetic rates.
   b. With the onset of dry period, subsequent leaves produced are smaller, and often with a pubescent covering. Lower photosynthetic rates, but lower heat load and transpiration rates.
   c. Examples: Brittlebush *(Encelia farinosa)* white bursage (*Ambrosia dumosa*).
4. Stem photosynthesis
   a. Stems lose less water than leaves (reduced transpiration).
   b. During period of low water stress, leaves are more productive than stems, but under periods of water stress, the superiority of leaves is lost.
   c. Example – cheeseweed (*Hymenoclea salsola*).
   d. A number of plants are essentially aphyllous (without leaves), relying on stem photosynthesis: Mormon tea

(*Ephedra spp.*), spiny menodora (*Menodora spinescens*), turpentine bush (*Thamnosma montana*).

Drought-enduring Plants

1. True xerophytes, maintaining a canopy and positive net carbon gain (photosynthesis) throughout the year.
2. Example: creosote bush (*Larrea tridentata*)
   a. Compound leaves (2 leaflets) commonly live for 8-14 months, but a canopy of photosynthetically active leaves is present throughout the year.
   b. Leaves of creosote are oriented more or less vertically, parallel to the sun.

c. Glandular trichomes secrete a resin that covers the leaf surface. The resin limits photosynthesis, but also drastically reduces transpiration.

3. Phreatophytes: These plants usually have extensive root systems which either spread through the surface soils or penetrate several metres below the surface Example – mesquite (*Prosopis*).
   a. Branching usually occurs in the capillary fringe above the water table in deep-rooted plants (phreatophytes) that are able to tap permanent ground water; while short-lived "rain roots" develop on woody surface roots in response to soil moistening.

Drought-resisting Plants – Succulents

1. Succulence is the most obvious characteristic of drought-resisting plants.
2. Desert succulents are generally shallow-rooted, allowing them to respond quickly to light rainfalls.
3. Stems are often heavily waxed to reduced cuticular water loss.
4. Leaves are often reduced to spines, and this increases the volume to surface ratio. In barrel cactus the ratio is 2.5, compared with 0.92 for a succulent leaf of an agave, and 0.01 for many non-succulent leaves. The maximum ratio is achieved by the spherical form of many cacti.
5. Spines also help to reduce heat load, and dissipate heat. Tissue temperatures below spines of the cholla cactus (*Opuntia bigelovii*) can be reduced by as much as 11 C.
6. Desert succulent are rarely killed by high temperatures, and several species of cacti and agave can withstand temperatures over 60C (140F) for short periods. However, their seedlings are

especially sensitive to high-temperature injury, and establishment is often prevented in open areas where soil temperatures can rise to 80 C (176F).

7. Seedlings of saguaro and other cacti require the shade of a nurse plant, like palo verde, to survive.
8. CAM plants – stomata of succulents open in the cool of the night. Transpirational water loss is reduced, restricting CO2 uptake to the dark hours. Assimilation products are stored at night, before being converted to photosynthates during the daylight period.
9. Water loss is very limited in desert succulents, but photosynthetic rates and, consequently, growth rates are low also. Annual growth rates in barrel cactus (*Ferocactus acanthodes*) is less than 2 cm.

Other Adaptations

1. Microphylly – typical leaves of warm desert plants are small and narrow, a design that enables leaf temperature to be near ambient temperature even when stomata are closed, so leaves can avoid lethal summertime temperatures during summer drought.
   a. Broad leaves are characteristic of plants in mesic environments or, if in the desert, in microclimatically mesic environments, like in a riparian zone, by a wash, or perched on top of a shallow water table, like the California palm (*Washingtonia filifera*).
2. Solar angles – Steeper leaf angles (parallel to solar radiation) reduces solar interception at midday during summer months, when air temperature and water loss may be most severe.
3. Solar tracking
4. Leaf reflectance / leaf absorptance
   a. Leaf absorptance in mesic communities is generally 85%.
   b. Leaf absorptance in desert communities range from 60-85%, but are as low as 29% in brittle bush (*Encelia farinosa*).
   c. Trichomes reduce heat load, reduce leaf temperature, reduce transpiration rates, reduces photosynthetically active radiation (negative effect), by absorbing and reflecting infra-red radiation.
5. Leaf type – Finely divided compound leaves, like those of Acacia, decrease the rate of water loss during periods of water stress, and reduce head load when stomata are closed.

What is Drought Resistance?

It is the ability of a plant to maintain favourable water balance and turgidity even exposed to drought conditions there by avoiding stress and its consequences. Stress avoidance due to morphological anatomical characteristics which themselves are the consequences of the physiological processes induced by drought these zerophytic characteristics are quantitative and vary according to environmental conditions. A favourable water balance under drought conditions can be achieved by transpiration before as soon as stress is experienced. These are called "water savers" or.

Accelerating water uptake sufficiently so as to replenish the lost water called as "water spenders"

The mechanism for conserving water:-

1. ***Stomatal mechanism:*** Stomata of different species vary widely in their normal behaviour and range. In some species stomata remain open continuously or remain closed continuously. Many cereals open their stomata only during a short time in the early morning and remain closed during rest of the day. There is a differencc in this respect between varieties of the same crop as shown by the example in two varieties of oat one is more resistant to drought open its stomata more rapidly in the early morning when moisture stress is at its minimum and photosynthesis can precede with the least loss of water (stocker 1960).

However mechanism of conserving water based on the closure of stomata will inevitable load to reduce photosynthesis and may lead to drought induced starvation injury (Leavitt, 1972).

2. ***Increased / Photosynthetic efficiency:*** On possibility for overcoming limitations on photosynthesis, imposed bicoastal closure as means for increasing resistance to loss of water by transpiration there by transpiration there by accumulations of CO2 would be at higher rate for a given stomatal opening (Hatch & stack, 1970). A number of imperfect crop plants (maize, sugarcane sorghum prose, fox tail & finger millets) (Hatch et. al. 1987) as well as certain forage species Bermuda grass (Cynodon dactyl on) Sudan grass Bahia grass (Paspalum notatum) Rhodes grass (chloris Guyana) (Murata lyama 1963) and certain A triplex sp. fixed most of CO2 into the C4 of molic and aspartic acids so called C4 dicarboxylic acid (C4) pathway.
3. ***Low rate of cuticular transpiration:*** The typical example is the cacturs. Thick cuticle results in low rate of transpiration.

4. Decreasing transpiration by a deposit of lipids layers on the surface of the leaves on exposure to moderate drought e.g. soybean (Levitt 1972).
5. ***Reduce leaf area:*** The principal means of reducing water loss of xenomorphic plants is their ability to reduce their transpiring surface. Apart from the common means of keeping the aerial parts small perhaps the simplest form of this reduction of the transpiring surface is the scaling or of leaves at the time of water stress a characteristic phenomenon exhibited by many grasses. The rolling of leaves has been shown to reduce transpiration by almost 55 percent in semi conditions and by 75 percent in desert xerophytes (Stalfect - 1956).
6. ***Leaf surface:*** Various morphological characteristics of leaves he reduce the transpiration rate and may affect survival of plants drought conditions. Leaves with thick cuticle waxy surface and the presence of spines etc. are common and effective.
7. ***Stomatal frequency and location:*** A smaller number of stomata retard the development of water deficits. In certain species, the stom are located in depression or cavity in the leaves which is feature can further reduce transpiration by limiting the impingement of currents.
8. ***Effect of awns:*** Awned varieties of wheat predominate in the drier at warmer regions and have been found to yield better than awnless one especially under drought conditions though there are exceptions (Gurandhacher 1963). Awns have chloroplasts stomata and so as photosynthesized. It has been found that the contribution of the away to the total dry weight matter of the kernels was 12% of that the entire plant.

To Improving water uptake (MC - Donough & Gauch 1959) :-

Efficient Root System: The root systems of drought resistant plants are characterized by wide variety of apparent adaptations. These responded to such predominant soil conditions as the duration of soil dryness and the depth that is normally wet. Plants become adapted to dry conditions mainly by developing an extensive root system rather that structural modification of the roots (shields - 1958). The conceres "extensive root system" includes additional growth of secondary hair roots.

High Root to Top Ratio (R/T): A high root to top ratio is very effective mean to adoption of plants to dry conditions of the growth rate of the roots considerably exceeds that of the shoots. The transpiring surface is there by reduced while root system of the individual plant obtains it's water from a large volume of soil (Simonis 1992) has shown that an increased root top ratio may actually result in greater amount of total dry matter of plants grown under dry conditions as compared a similar ones grown with full moisture.

Difference in Osmotic Potential of Plants : Levitt (1958) has calculated a difference of 0.5% in soil moisture content that includes per manual wilting could supply a plant with enough water to keep it alive for 6 days. This could mean in certain cases the difference between survival and death.

Conservation of Water Spenders to Water Stress: Because of increased water absorption water spenders are characterized by very high rate transpiration. However as soon as the absorption rate becomes insufficient to keep up with water loss the water spenders generally develop some of the characteristics of the water savers (Cevitts - 1972).

Mitigating Stress

Mitigating Stress: Adoptions a drought basis mitigating effects of stress permit the plant to maintain a high internal water potential inspite of drought conditions. They therefore able to maintain cell tartar and growth avoid direct or indirect metabolic injury due to dehydration (Levit 1972).

Drought Tolerance

When plant is actually submitted to low water potential it can show drought tolerance by either mitigating the actual stress induced by the moisture deficiencies or by showing high degree of tolerance to stresses.

High Degree Tolerance; Resistance to Dehydration: The simplest method of avoiding drought induced damage is by resisting dehydration, preferably tot he extent .of maintaining turgur and at least by avoiding cell collapse after loss of turgur (Levit 1972) retain their turgur and therefore can continue to grow when exposed to drought stress. When plants are grown in their natural environment their osmotic potentials tend to be characteristic for each ecological group.

Agro Climatic Zones of India In General

Introductions: - The important rational planning for effective land use to promote efficient is well recognized. The ever increasing need for food to support growing population @2.1% (1860 millions) in the country demand a systematic appraisal of our soil and climatic resources to recast effective land use plan. Since the soils and climatic conditions of a region largely determine the cropping pattern and crop yields. Reliable information on agro ecological regions homogeneity in soil site conditions is the basic to maximize agricultural production on sustainable basis.

This kind of systematic approach may help the country in planning and optimizing land use and preserving soils, environment.

India exhibits a variety of land scopes and climatic conditions those are reflected in the evolution of different soils and vegetation. These also exists a significant relationship among the soils, land form climate and vegetation. The object of present study is to delianate such regions as uniform as possible introspect of physiographic, climate, length of growing period (LPG) and soils for macro level and land use planning and effective transfer of agro - technology.

Agro Climatic Zones: - Agro climatic zone is a land unit in Irens of mator climate and growing period which is climatmenally suitable for a certain image of crops and cultivars (FAO 1983). An ecological region is characterized by district ecological responses to macro - climatic as expressed in vegetation and reflected fauna and equatic systems. Therefore an agro-ecological region is the land unit on the earth surface covered out of agro - climatic region, which it is super imposed on land form and the kinds of soils and soil conditions those act as modifiers of climate and LGP (Length of growing period).

With in a broad agro climatic region local conditions may result in several agro - ecosystems, each with it's own environmental conditions. However, similar agro ecosystems may develop on comparable soil, and landscape positions. Thus a small variation in climate may not result in different ecosystems, but a pronounced difference is seen when expressed in vegetation and reflected in soils.

India has been divided into 24 agro - climatic zone by Krishnan and Mukhtar Sing, in 1972 by using "Thornthwait indices".

The planning commission, as a result of mid. term appairasal of planning targets of VII plan (1985 - 90) divided the country into 15 broad agro - climatic zones based on physiographic and climate. The emphasis was given on the development of resources and their optimum utilization in a suitable manner with in the frame work of resource constraints and potentials of each region. (Khanna 1989).

Agro climatic zones of India :- (Planning Commission 1989)

1	Western Himalayan Region	Ladakh, Kashmir, Punjab, Jammu etc.brown soils & silty loam, steep slopes.
2	Eastern Himalayan Region	Arunachal Pradesh, Sikkim and Darjeeling. Manipur etc. High rainfall and high forest covers heavy soil erosion, Floods.
3	Lower Gangatic plants Regions	West Bengal Soils mostly alluvial & are prone to floods.
4	Middle Gangatic plans Region	Bihar, Uttar Pradesh, High rainfall 39% irrigation, cropping intensity 142%
5	Upper Gangatic Plains Region	North region of U.P. (32 dists) irrigated by canal & tube wells good ground water
6	Trans Gangatic plains Region	Punjab Haryana Union territory of Delhi, Highest sown area irrigated high
7	Eastern Plateaus & Hills Region	Chota Nagpur, Garhjat hills, M.P, W. Banghelkhand plateau, Orissa, soils Shallow to medium sloppy, undulating Irrigation tank & tube wells.
8	Central Plateau & hills Region	M. Pradesh
9	Western Plateau & hills Region	Sahyadry, M.S. M.P. Rainfall 904 mm Sown area 65% forest 11% irrigation 12.4%
10	Southern Plateau & Hills Region	T. Nadu, Andhra Pradesh, Karnataka, Typically semi and zone, Dry land Farming 81% Cropping Intensity 11%
11	East coast plains & hills Region	Tamil Nadu, Andhra Pradesh Orissa, Soils, alluvial, coastal sand, Irrigation
12	West coast plains & Hills Region	Sourashtra, Maharashtra, Goa, Karnataka, T. Nadu, Variety of cropping Pattern, rainfall & soil types.
13	Gujarat plains & Hills Region	Gujarat (19 dists) Low rainfall arid zone. Irrigation 32% well and tube wells.
14	Western Dry Region	Rajasthan (9 dists) Hot. Sandy desert rainfall erratic, high evaporation. Scanty vegetation, femine draughts.
15	The Island Region	Eastern Andaman, Nikobar, Western Laksh dweep. Typical equatorial, rainfall 3000 mm (9 months) forest zone undulating.

Rainfall, Its Distribution and Its Effectiveness In Rainfed Agriculture

Distribution of Rainfall: - The amount of rainfall received at periodic intervals like weeks, months, seasons etc. indicate distribution. In addition distribution of rainfall can be known by the length of dry spell, wet spells land rainy days. Distribution of rainfall is more important than total rainfall.

Rainfall pattern at three locations:-

Sr. No.	*Index*	*Hyderabad*	*Solapur*	*Dhule*
1	Annual rainfall (mm)	764	742	625
2	Seasonal rainfall (mm)	580	556	450
3	Coefficient of variation%			
4	PE (mm)	1757	1802	1502
5	Growing season	130	148	125
6	Soil	Vertisols	Vertisols	Medium deep

From the above data the growing season is slightly less at Hyderabad as compared to Solapur, However rainy season crops are more successful at Hyderabad and annual yields range 50 to 70 q/ha. While at Solapur and Dhule rainy season crops are risky and annual yields range from 10 to 12 q/ha. Low yields at Solapur and Dhule are mainly due to discontinuous rains or long breaks in rainfall during crop growth period.

Rainfall distribution is based on:

1. Weekly or monthly rainfall will give distribution of rainfall in weeks during a crop season.
2. Wet and dry spells - A wet spell is a number of continuous days of rainfall. A dry spell is a number od continuous rainless days.
3. Rainy days: If the rainfall received is more than 2.5 mm on any day. This particular day is called rainy day.
4. Periodicity of rainfall.
5. Onset of monsoon.
6. Recurrence of rainfall events.
7. Dependability of rainfall
8. Certificient of variation. If C.V. is more variation in rainfall is more and vice - a - versa.
9. Length of the growing season (LGS): If LGS is less a short duration crop should be selected. L.G.S. depends on duration of rainy season and moisture retention.

Rainfall being a single most important factor for success of crops in the dry farming areas. It is generally known that India receives its annual rainfall by the particular phenomenon called monsoon which consists of series of cyclones those arise in the Indian Ocean. These travel in the North East direction and enter the peninsular India along the Western coast. These cyclones occur from June to Sept. is known as south West monsoon. This is followed by second third and fourth rainy season during periods from Oct. to Nov., Dec. To Feb. and March to May respectively. South West Monsoon is the most important as it covers major parts of India and brings bulk of the total annual rainfall.

The North East of Returning monsoon: By the end of Sept. South West Monsoon ceases to penetrate North West India but continues a full month longer in Bengal. On account of south East North easterly winds being to flow on the Eastern coast. Some times some of these cyclones penetrate In land and give supplementary rainfall to dry region of the plateau of the peninsular India. This is known as returning monsoon.

Precipitation and Its Factors

Precipitation is reaching of atmospheric humidity either as rain or snow to the ground. OR Precipitation can be defined as earth word falling of water drops of ice particles that have formed by rapid condensation in the atmosphere and are too large to remain suspended in the atmosphere.

Factors influencing precipitation:-

1. Only blowing of winds coming even over the sea is not enough to produce precipitation.
2. Horizontal movement is not conductive to precipitation.
3. The rain bearing clouds, hills, mountains, slanting slopes of the river ralleys lake dynamic cooling of the clouds.
4. Water vapour in atmosphere and moat conditions which promote greater precipitation.
5. Regiour covered with thick forest contributes more water vapour by transportation and thus provide favourable condition for precipion.
6. The prevalence of dry winds, higher temperature absence of barriers and cutting of monsoon currents these are unfavourable for precipitation.
7. Long & short breaks in the monsoon caused due to prevalence of dry winds slowing over land or desert plains from North East.

8. On the other hand geographical position, physical configuration and meteorological conditions are responsible for precipitation.

Type of Rainfall In Dry Areas

1. First type rainfall: Rainfall receives from south west Monsoon. Rainfall receives up to 60% in the first three months viz. June - July - August - Rontak Jodhpur Jalgaon.
2. Second type rainfall: Rainfall receives from south - West Monsoon (40 to 55%) and supplemented with North East Monsoon (40 to 50%) Pune. Wal A Nagar Raichur - Maximum rainfall receives in July & Sept.
3. Third type rainfall: Rainfall receives from North East Monsoon (60%) Solapur Bijapur a Karnataka.
4. Four type rainfall: Rainfall receives uniformly (Well distributed) from Both Monsoon currents, places of rainfall - Chennai. Total rainfall received in 5 to 6 months.

Decennial rainfall: The mean total rainfall received during past 10 years:

Winds coming over land surface from the North - East are dry and cold and man cause of breaks in the monsoon or they tend to decrease the rainful of a tract by diluting the moisture laden masses of the atmosphere. Indian be divided into three zones of the basis of rainfall.

A) Heavy rainfall zone: above 1250 mm.

B) Moa crate rainfall zone: 750 to 1250 mm.

C) We rainfall zone: Less than 750 mm annual rainfall.

The average rainfall of Solapur varies from 500 to 720 mm and had bimodal stribution. The first peak is usually experienced during June and Second during Sept. Rainfall during Sept. is more assured and is in the rage of 150 to 200 mm. Even though Monsoon sets in by the end of June, July and August are characterized by dry spells of varying duration (2 to 8 weeks at stretch) and frequencies 1 to 5. Usually dry speels of more that 4 weeks duration or 3 dry spells of 2 week duration result in failure off Kharif crops. Such occasions are observed twice in fire years. Usually high wind velocity (18 to 20 km / hr)

At Solapur under dry land areas year to year fluctuatins are so much that there is no guarantee of a fixed quantity of rainfall. Generally rainfall starts in late June to early July. There is depression during late July to early August. Again there is good amount of rainfall in last Aug. and Sept. The rainfall totally recedes by mid October. This is the usual pattern of rainfall in draught prone areas. The probability of rainfall is more than half the normal is fairly good. (P = 0.58) during September.

Techniques of Soil and Water Conservation In Rainfed Agriculture

Soil and water are most essential for the growth and sustenance of plant life. Soil is important as it provides, foothold for plants and majority of nutrients needed by them. Water is essential as it forms larger part of the living matter and acts as a nutrient carrier.

Though, both soil and water a sources are available in plenty, they are not distributed equally in quality and quantity in every part of the world and are not inexhaustible. Their abuse would mean a great loss resulting in poverty. It takes centuries to form one inch layer of soil, but it does not take long to lose it by erosion. Research work carried out in Maharashtra State which has an undulating topography, has shown that loss of soil from unprotected land is as much as 125 tons per hectare every year and may be as high as 300 tons in a single year. The weight of one hectare of soil 2.5 cm. deep is about 325 tons.

Similarly, rain water which can sustain a good crop, if not conserved properly will not only cause scarcity and famine, but also wash way the soil which is a valuable national asset. There are many examples which show how once fertile plains and valleys have become deserts or barren lands due to neglect by mankind. It is, therefore, the prime responsibility of each generation to conserve soil which is the main capital of the farmer as well as the nation, at all costs and pass it on in good condition from one generation to another, so that the posterity will not blame them.

Soil and water conservation cannot be achieved only by individual efforts. The problem is too big, involving collective efforts on the part of farmers, technicians and Government. Recognizing the seriousness of erosion problem, the Central Government established the Central Board of Soil Conservation to assist the States and River valley Projects. It has established soil conservation research stations at Dehradun, Kotah, Ootacamand, Bellary, Vasad and Jodhpur, arranges for training of technical personnel and also served as clearing house for soil conservation information.

On account of chronic scarcity conditions prevailing over 3 major portions of the Deccan tract. Soil and water conservation research was started in 1924 and soil conservation work was taken up on a large scale in Maharashtra from 1943 - 44 onwards. At present Maharashtra contributes nearly 50 p.c., of the total progress in respect of soil conservation measures in the country.

Water Conservation

Along with soil water is another important factor essential for all life and production of food. The main source of water is precipitation. In India precipitation is not property distributed throughout the year. It is received within a few months of rainy season and that too in a critic manner. It may rain 50 to 125 mm. in one day causing flood and damage to crops and then there may be a dry spell for some days when crops may begin to will. Hence proper conservation of water as well as collecting surplus water in tanks and reservoirs of letting it out into the rivers assume great importance. To understand water conservation it is necessary to study the hydrologic cycle.

Hydrologic Cycle

Water moves in a continuous cycle from ocean to clouds to earth and back to ocean. The water in the ocean is converted into vapour by the heat of the sun and this vapour moves in the form of clouds over the land and condenses into rain. Some rainwater enters the soil while the rest flows into streams and rivers and is either stored in tanks and reservoirs or allowed to go back to sea. Of the water that enters the soil some is stored for use by the plant some gets evaporated from the surface of the soil and some moves down to replenish the water table. This becomes the source of water for wells and springs. This cyclic movement of water is known as the hydrologic cycle. The water that is held by the soil in available form is essential for plant growth. It is not yet possible for the man to control rainfall but its infiltration and run off can be regulated to a great extent by improved management practices. Efforts should be made to store rain water either in the soil or in the reservoirs when it is in plenty and carry over to periods when rain water is not available. The former is known as conservation of water in soil, while the latter, conservation in reservoirs.

Loss of Water From Soil

Water is lost from soil in four ways:

1. Surface run off.
2. Downward movement of drainage.
3. Evaporation from soil surface.
4. Transpiration through leaves of plants.

Out of this loss of water through run off is the largest and is also the most damaging as it causes erosion main factor in conserving moisture relates to increasing infiltration and storage capacity of soil and reducing run off and evaporation. Uncontrolled water is the main cause of soil erosion. Almost all methods that deal with soil conservation are in principle the methods to control and conserve water. Soil and water conservation are, therefore, dealt together, Sufficient studies on all phases of hydrologic cycle such as evaporation, precipitation, run off, infiltration and deep percolation which occur simultaneously have not been carried out in India as yet, though 2 beginning has been made at many agricultural research stations.

Soil and Water Conservation Methods

The loss of soil and water under natural vegetation is the lowest. But lands must be cultivated and grown with crops to produce food. This can be done without much harm to the soil if proper soil and water conservation methods are followed. Such methods aim at encouraging water to infiltrate into the soil, reduce its velocity and check run off losses.

The most common soil and water conservation methods are

A) Management practice viz.
   a) Strip cropping,
   b) Mulching,
   c) Crop rotation,
   d) Contour cultivation,
   e) Planting of grasses for stabilizing bunds,
   f) Planting of trees and a forestation,
   g) Cashew nut plantation, and

B) Mechanical practices such as
   a) Bunding,
   b) Terracing,
   c) Gully or nala control,
   d) Control of stream and river banks.

Soil and Water Conservation Methods - Management Practices

Strip Cropping: This consists of growing erosion permitting crops and erosion resisting crops in alternate strips. The erosion permitting crops are cotton, jawar, bajara, etc. which allow the run off water to flow freely within the rows. The erosions resisting crops are mostly legumes like groundnut, much (Phaseolus aconitiolius), hulgn (Dolichos biflorus), Sow bear (Glycine max) which spread and cover the soil and do not allow run off water to carry much soil with it the soil which flows from the strips growing erosion permiuming crops is caught by the alternating springs.

In selecting a suitable legume crop it should be seen that the maximum canopy and root development of the crop coincide with the period of high intensity of rainfall.

Mulching: A mulch is natural or artificially applied layer of plant residues or other material on the surface of the soil with the object of moisture conservation, temperature control, prevention of surface compaction or crust formation, reduction of run off and erosion, improvement in soil structure and weed control. Artificial mulches of different kinds such as *Jowar* or *bajara* stubbles, stubbles, paddy straw or husk, sawdust etc., increase absorption of water and minimize evaporation. They also control run off and soil losses.

Rotation of Crops: Rotation means growing a set of crops in a regular succession over the same field within a specified period of time. Continuous growing of Jowar or bajara crop causes more erosion, but if followed by a legume crop viz., hulga, matki or gram which covers the soil is causes less erosion. Rotation also helps in removal of plant nutrients in a uniform way from future depth of soil and in maintaining the fertility of the soil in dry farming region of Maharashtra adoption of gram Jowar rotation not only helps in conservation of moisture but also in increasing the crop yields the beneficial effect of rotation.

Contour Cultivation: Tillage operations viz., ploughing, harrowing, sowing and Interculture should be done across the slope of land. This will help in creating obstructions to the flow of water at every furrow, which acts like a small bund and results in uniform distribution of water. This helps more initration of water less run off and erosion, and gives higher crop yield. Any cultivation done along the slope will accelerate golly formation, more run off and erosion and consequently permanent damage to land.

Planting of Grasses for Stabilizing Bunds: Grasses prevent soil erosion and improve soil structure. The entire soil mass is penetrated by countless roots and soil aggregates and particles are enmeshed by the root system. Grasses should be grown on bunds which are not suitable for cultivation, both for checking erosion and providing pasture for cattle. Several grasses as well as legumes were tried on bunds at the Agricultural Research Station; Solapur which receives about 600 mm. of rainfall to see which of them will withstand drought conditions, give maximum root growth and canopy coverage, and stabilize bunds effectively. It was observed that anjan planted with spacing of 15 x 15 cm., produced the highest quantity of roots, followed by marvel - 8, Rhodes, thin Napier, blue panic, and *kusal* .Legumes do not have many roots but produce better canopy within a short period, while grasses are under the process of establishment. Planting of legumes mixed with grasses is, therefore, advantageous in preventing soil erosion in initial stages.

Planting of Trees and Afforestation: Forests conserve soil and water quite effectively. They not only obstruct the flow of water, but the falling leaves provide organic matter which increases the water holding capacity of the soil. If tree planting is done in the planned manner in open areas, it will serve as good wind break and if done along the banks of streams and rivers, it will regulate their flow. Farm forestry is another important aspect in soil and water conservation. The danger caused by deforestation has been only recently appreciated and a big plantation programme of hybrid eucalyptus, teak, casurina has been taken up by the Forest Departments in reserve forests, catchments areas of irrigation projects and on Government waste lands Vanamahotsava is also observed every year in the early part of monsoon and millions of trees are planted by the public with the help of the maff of the Agricultural and Forest Departments local bodies like Zillah Paris had, Panchayat Samitis and Gram Panchayat What is however important is to pay proper attention planted.

Cashew Ant Plantation : In coastal districts of Maharashtra which receive more than, 1,250 mm. rainfall a new programme of cahsewnut plantation has been undertaken from 1963 - 64 on hills having slope between 10 and 20 p.c.The sea breeze is conducive to the growth of cahsewnut plants and they do not require much aftercare once they establish in the soil. Staggered trenches of 300 x 30 x 30 cm. size are dug on contours at a distance of 6m. Cashew plants are raised in polythene bagsin nurseries and two months olds a plings are planted on the lower side of the trench with plant to plant distance of 6m.

Soil and Water Conservation Methods - Mechanical Practices

The above measures control erosion by good management practices. Bunding, terracing, *gully* or *nala* control, and construction of tanks and bandharas are mechanical measures requiring engineering techniques and structures. They reduce run off and impound water for longer time to help infiltration into the soil. Their construction and design will depend upon rainfall, soil slope and such other factors. These measures are costly but if properly maintamed will improve the land over a long period of time.

Bunding

Block Bunding: Bunding for control of soil erosion and conservation of surface run off was known to farmers for centuries. It was not uncommon to find Tals i.e., big bunds across large blocks of sloping lands. These bunds are constructed of earth or stone or both, at a great cost, to impound water and arrest soil washed from the fields lying above. They are high and broad enough to withstand the force of water from the catchments. Water is let out at the end of the monsoon and land which has received fertile silt is sown with crops. Such type of big bloods bunds are not constructed now as contour bunding has been taken up on catchments basis.

Contour Bunding: It consists of construction of a serics of earthen bunds of suitable sizes along contours at a lateral distance of every 60 mm or a fall of 1 to 1.5 m. The shope of land is thus broken into smaller and more level compartments which hold soil as well as rain water. It has been estimated that about 75 million hectares of land i.e. about one fourth of the common land surfaces suffering from soil erosion. In Maharashtra State, the problem is more acute and it is estimated that out of 186 lakh hectares about 144 lakh hectares require bunding. The planning Commission has, therefore laid great stress on contour bunding programme, because bunding alone has been found to increase crop yield by 20 to 30 p.c.

The size, cross- section and interbund spacing depend upon the nature of rainfall, soil and slope of the area. In order to improve the technique of bunding, studies have been carried out in respect of spacing of bunds, shrinkage of bund sections and hydraulic gradients and kind and location of outlets etc. in different soil and rainfall conditions of Maharashtra State. On the basis of such studies it has been observed that the spacing between bunds should not be allowed to exceed 1.5m. Vertical drop or 67.5 m. lateral spacing.

Graded Bunding: In high rainfall areas, while conservation of soil is important, drainage of surplus water has to be attended to, for avoiding waterlogged condition of soil. The bunds are therefore, slightly graded longitudinally about 7.5 cm. per running 33 m. to prevent safe disposal of water into the nala. The cross sections into for safe removal of excess run off water it is essential to provide suitable waste water or outlet structures at proper places so that no damage is done to bunds in case heavy precipitation is received on any single day. Normally stone outlets are provided low rainfalls are as. Channel weirsor pipe outlets may also be provided. Grass outlets have been found to be effective and cheaper in heavy soil. The crest wall should be 30 cm. above the contour level and its length should be so designed as to discharge the surplus water from the maximum intensity of rainfall with are asonable period. 1,250 mm. Terrace bunds consist of comparatively narrow embankments constructed at intervals across the shope and the vertical spacing between bunds may vary from 1to 2 m., depending upon the slope, type of soil, rainfall etc. Bench terracing is done when gradient is stceper than 10 p.c. as in hilly ranges of Himalayas, Sahyadry etc. and consists of a series of step like platforms along contours. These terraces are like table tops sloping outwards and are provided with stone wateweirs to drain away surplus water. Angular and big boulders should be used for terrace outlets because round and small boulders will slip and get dislodged under the gushing water.

Gully or *nala* control: Gully or nala control is very essential to prevent its extension and further destruction of cultivated lands and grasslands. The sloping sides are planted with grass and trees. Suitable temporary and permanent structures such as check dams, overflow dams, drop structures are also provided. Small gullies can be stabilized by converting them into paddy fields. So far 17,034 nalas have been controlled and the target for sixth plan (1980 - 85) period is 2005.

Control of stream and river banks: Vulnerable sharp bends nalas by the sides of the rods and river bends near village sites cause considerable damage to property. These should be protected by providing spurs, jetties, rivets and retaining walls. Adjoining areas should be stabilized under permanent vegetation. Spurs are constructed at an angle to reduce the velocity of water and there by enabling the flood water to flow away but deposit coarse sand which will cause obstruction to successive water currents from cutting into the bank and thus straightening their course.

Runoff, and Their Types

Definition:" It is that portion of rainfall, which makes its way towards streams, rivers etc. After satisfying the initial losses etc. is called as runoff.

Types of Runoff:

1. Surface runoff.
2. Sub - surface runoff, and
3. Base flow.

1. Surface Runoff: It is that portion of rainfall which enters the stream immediately after the rainfall. It occurs. When all losses are satisfied and if rain is still continued, with the rate greater than in filtration rate; at this stage the excess water makes a head over the ground surface (surface detention) which tends to move from one place to another, known as overland flow. As soon as the overland flow joins to the streams, channels or oceans, termed as surface runoff.
2. Sub - surface Runoff: That part of rainfall, which first leaches into the soil and moves laterally without joining the water -

table to the Streams Rivers or oceans is known as sub - surface runoff. Sometimes sub - surface runoff is also aerated under service ninoff due to reason that it takes very title time to reach the river or channel in comparision to ground water. The sub - surface runoff is usually referred as interflow.

3. Base flow: It is delays flow, defined as that part of rainfall which after talling on the ground surface in fill rated into the soil and meets so the water table and flow to the streams oceans etc. The movement of water in this type of runoff is very slow that is why it is also referred as delayed runoff. It takes a long time to join the rivers or oceans. Some times base flow is also known as ground water flow.

Thus,

Total Runoff = Surface runoff + Base flow (Including sub - surface runoff)

Factors Affecting Runoff

The runoff rate and its volume from an area, mainly in influenced by following tow factors:

a) Climatic factors and

b) Physiographic factors

Factors Affecting Runoff - Climatic Factors:

The climatic factors of the watershed affecting the runoff are mainly associated with the characteristics of precipitation, which include.

1. Type of precipitation
2. Rainfall intensity
3. Forms of precipitation
4. Duration of rainfall
5. Rainfall distribution
6. Direction of prevailing wind and
7. Other climatic factors.

Type of Precipitation: Types of precipitation have a great effect on the runoff. For example a precipitation which occurs in form of rainfall, starts immediately in from of surface flow over the land surface, depending upon its intensity as well as magnitude, while a precipitation which takes place in form of snow or hails the flow of water on ground surface will not take place immediately, but after melting of the same. During the time interval of their melting the melted water infiltrates into the soil and results a very little surface runoff generation.

Rainfall Intensity: The intensity of rainfall has a dominating effect on runoff yield. If rainfall intensity is greater than infiltration rate of the soil the surface runoff takes place very shortly while in case of low intensity rainfall, where is found a reverse trend of the same. Thus high intensities rainfall yield higher runoff and vice-versa.

Duration of Rainfall: Rainfall duration is directly related to the volume of runoff due to the fact, that infiltration rate of the soil goeson decreasing with the duration of rainfall till it attains constant rate. As a result of this even a mild intensity rainfall lasting for longer duration may yield a coverderable amount of runoff.

Rainfall Distribution: Runoff them a water sheed depends very much on the distribution of rainfall. The rainfall distribution for this purpose can but expressed by a team "distribution coefficient which may be defined as the ratio of maximum rainfall at a point to the mean rainfall of the watershed. For a given total rainfall, if all other conditions are the same, the greater the value of distribution coefficient, greater will be the peak runoff and vice - versa. However, for the same distribution coefficient, the peak runoff would be resulted from the storm, falling on the lower part of the basin i.e. near the outlet.

Direction of Prevailing Wind: The direction of prevailing wind, affected greatly the runoff flow. If the direction of prevailing wind is same as the drainage system then it has great influence on the resulting peak flow and also on the duration of surface flow, to reach at the outlet. A storm moving in the direction of stream slope produces a higher peak in shorter period of time than a storm moving in opposite direction.

Other Climatic Factors: The other climatic factors, such as temperature wind velocity, relative humidity, annual rainfall etc. affect the water losses from the watershed area to a great extent and thus the runoff is also affecter accordingly. If the losses are more the runoff will be less and vice -versa.

Factors Affecting Runoff - Physiographic Factors

Physiographic Factors: Physiographic factors of watershed consist of both, the watershed as with as channel characteristics. The different characteristics of watershed and channel, which affect the runoff, are listed below.

1. Size of watershed
2. Shape of watershed
3. Slope of watershed
4. Orientation of watershed

5. Land use
6. Soil moisture
7. Soil type
8. Topographic characteristics, and
9. Drainage Density.

Size of Watershed: Regarding the size of watershed, if all other factor including depth and intensity of rainfall are being same them two watershed irrespective of their size, will produce about the same amount of runoff .However a large watershed takes longer time for raining the runoff to the outlet as result the peak flow expressed are depth is being smaller and vise versa.

Shape of Watershed:The shape of watershed has a great effect of runoff. The watershed shape is generally expressed by the terms "from factor and "compactness coefficient".

Shope of Watershed: The shope of the watershed has an important roel over runoff but its effect is complex. It controls the time of overland flow and time of concentration of rainfall in the drainage channel which provide accumulative effect on resulting peak runoff. For example in case of a sloppy watershed. The time to reach the flow at outlet is less, because of greater runoff velocity which results into formation of peak runoff very soon and vice -versa.

Orientation of Watershed: This factor affects the evaporation and transpiration losses from the area by making influence on the amount of heat to the received from the sun. The north or south orientation of watershed, affects the time of melting of collected snow. In a mountainous watershed the part of wind ward side of the mountain receives high intensity of rainfall resulting into more runoff yield while the part of watershed typing towards leeward side has reverse find of the same.

Land Use: The land use pattern and land management practices used have great effect on the runoff yield. For example an area which is under forest cover, where a thick layer of much of leaves and grasses etc. has peen accumulated there formed a little surface runoff due to the fact that more rain water is absorbed by the soil. While in a barren field where not any type of cover is available a reverse trend is obtained.

Soil Moisture: The magnitude of runoff yield depends on the amount of moisture present in the soil at the time of rainfall. If rain occurs over the soil which has more moisture the infiltration rate becomes very less which results in more runoff yield. Similarly if the rain occurs after a long dry spell of time when the soil is dry, causing to absorb huge amount of rain water. In on the other hand, if the rain occurs in a close succession as in the rainy season; runoff yield has reverse effect.

Soil Type: In the watershed surface runoff is greatly influenced by the soil type as loose of water from the soil is very much dependent on inflientation rate which varies with the types of soil.

Topographic Characteristics: Topographic characteristics include mores topographical features of watershed which create their effect on runoff it is mainly undulating nature of the reason that runoff water gets additional power to flow due to slope of the surface and altitude time to infiltrate the water into solid.

Regarding channel characteristics to describe their effect on runoff the channel cross-section, roughness storage and channel density are mainly considered. These also have significant effect on runoff.

Drainage Density: The rain age density is defined as the nation of the trial channel length in the watershed to two total watershed areas it is expressed at.

$$\text{Drainage density} = \frac{\text{(Tranned length (Total))}}{\text{Watered area}}$$

$$\text{D.D.} = \frac{\text{I}}{\text{A}}$$

A watershed having greater D.D. and incites formation of peak rain off very shortly to that of lesser D.D. watershed.

Different Agronomical Practices for Soil and Water Conservation

Conservation In Rainfed Areas

Soil conservation is a preservation technique, in which deterioration of soil and its losses are conserved by using it within its capabilities and applying conservation techniques for protection as well as improvement of soil. In hilly regions. Where land topography has steep slope and is subjected to erosion problem the vegetation cannot get established. Lack of the vegetative cover on sloppy soil surface accelerates the erosion and a large amount of soil is transported into the stream through runoff. In addition, the uncovered sloppy land also a cause extensive damage to the cultivable land at foothill through exporsition of sedimentson them.Sediment disposition covers the top fertilesoil layerand thus makes them unsuitable for cultivation.

Under this circumstance it becomes very necessary to treat such areas by adopting appropriate agronomical measures, so that they can be reclothed with negetations. The vegetation helps in reducing the surface runoff and soil cravsion both. The agronomical measures include contouring strip cropping and niluge practices to control they soil erosion. The use of these measures is entirely dependent upon the soil types land shope and rainfall characteristics.

In soil and water conservation programmes, thc agronomical practices are counted as second line of defence the first being mechanical or engineering measures which are employed to arrest the soil erosion immediately. The role of agronomic measure is more economical long-lasting and effective. Always it is advisable to used but when its use is either inadequate or not sulpewant to achieve the goal of erosion control then use of mechanical measures to control erosion is recommended.

The agronomical measures are referred by the practices of growing vegetables on mild sloppy lanks to cover them and to control the erosion from there in living vegetation above the soil surface dissipates the crove power of agents either they are water or wind In case of water erosion it affects by several ways such as by enhancing infiltration rate and relucing together and thereby reducing runoff velocity to scour the soil particles screening the eroded particles to reach them into the channels or reservoirs; by dissipating the kinetic energy of falling raindrops and thus reducing the splash erosion. The effect of vegetation on wind erosion is also significant as it directly makes a hinderance in blowing path and thus deflecting the wind current at some distance away towards down stream side. The wind - strip cropping is a well known agronomical practice comployed for controlling the wind erosion in wind erosion susceptible areas.

The role of agronomical measures in achieve of soil & water conservation, has immense importance, perhaps much more than the others. It can be explained by considering the Universal Soil Loss Equation (A = R K L S C P) in which agronomical practices reflect the factor of crop management (C). The other factors such as R & K are the natural factor; we do not have any control on them. The L S and P factors may have value as I under worst conditions; although these can be reduced maximum up to 0.5 by applying an ideal soil and water conservation measures. The factor 'C' which is crop management factor has value as I for worst conditions, but it can be reduced up to 0.02. At this small value of C, the soil loss can be minimized up to one - fifteenth which is about 10.25 times more than the other factors. Looking this important effect of agronomical measures on soil loss, its scope is assumed to be more dominating in soil and water conservation programmes.

1. Contouring
2. Trip Cropping
3. Tillage Practices

These are the important agronomical practices employed for controlling the soil erosion from sloppy areas. Basically these measures create an obstruction in flow path of surface runoff by making the land surfaces rough due to channels ridges etc. formed under them. Each of these measures also have a direct relation with the infiltration rate and thereby presence of moisture in the soil profile. Infiltration rate is an effective factor in reducing the surface runoff and soil loss.

Chapter 6

Fertilizer Use in Dryland Farming

To meet the increasing food grain requirement of the country, dry lands will have to playa crucial role. A large chunk of country's poor population inhabits the dryland areas. An attempt to improve the productivity of dryland areas intend to serve two purposes: improving the economic status of poor masses deriving their livelihood from dryland areas and contributing to the total food. grain production of the country. Fertilizers have a key role to play in improving productivity of dryland areas. To ensure the benefit and minimize the risk, a proper understanding of certain theoretical and practical aspects of fertilizer application in dryland areas is necessary.

Nearly 70 % of the arable land in our country is rainfed. This large chunk of area contributes only 42 % of the total food grain production. The scarcity of water in these areas greatly inhibits adoption of modern package of practices in farming including the use of high yielding varieties, pesticides and fertilizers. Consumption of fertilizers per hectare in dryland regions is nearly 3 times less than that in irrigated areas. More than 80 % of the total fertilizers used in the country is dumped in only 30% irrigated area. Researches have, however, proved that the productivity of dry lands can be improved significantly by adopting appropriate method of fertilizer application.

Dry lands have not only a thirst for water but also a hunger for nutrients. Nitrogen deficiency is a widespread problem of dryland areas. Low amount of moisture in the soil and relatively higher temperature encourage the loss of nitrogen from the soil. Soil nitrogen is lost to the atmosphere in gaseous form. Deficiency of phosphorus, potassium, sulphur and zinc is also frequently observed in dryland areas. These nutrients are essential for the growth and development of plants.

Nutrient Availability

Nutrient availability refers to the amount of nutrients present in the soil which plants can easily absorb by their roots. A nutrient is available, only when it is soluble in water. In the regions of adequate rainfall/irrigation, main thrust is to make the nutrients more and more soluble in the water, so that the plant roots may absorb them. But in dryland areas problem is just contrast. These soils do not have enough moisture to dissolve plant nutrient, even if the nutrient is present in the soil in water soluble form. Therefore, nutrients in the soil do not find their way into the plant roots. Application of fertilizers is, therefore, highly restricted in dryland areas.

Recommendation of Nutrients

Out of three primary fertilizer nutrients namely I nitrogen, phosphorus and potassium, nitrogen has the highest bearing on the crop yielding in dry land farming. This is because of the fact that dryland soils are highly deficient in nitrogen. Response of nitrogenous fertilizers is always greater than phosphatic and potassic fertilizers. On the other hand more amount of nitrogen can be made available than phosphorus and potash at a definite level of soil moisture. Nitrogen is more mobile in the soil than phosphorus and potash. A definite amount of moisture can transport more amount of nitrogen than phosphorus and potash. Yet, the amount of nitrogen recommended for crops in dryland condition is always less than in irrigated conditions. Generally less amount of fertilizers is recommended for cereal crops in dryland condition than in irrigated conditions. But for leguminous crops there is not much difference. Recommendation of doses for main crops grown in dryland areas.

Table: *Nutrient requirements of various crops in various regions.*

Sl No.	*Region*	*Crop*	*Quantity of nutrient kg/ha*		
			N	P	K
1	2	3	6	5	6
1.	Kharif black soils of Jhansi(U.P.)	Sorghum	0	40	0
		Red-gram	0	40	0
		Sesamum	0	30	0
		Guar	0	60	0
		Groundnut	0	25	0
		Wheat	0	30	0
		Barley	0	30	0
		Linseed	0	20	0
		Mustard	0	20	0
		Safflower	0	20	0

Contd...

Sl No.	*Region*	*Crop*	*Quantity of nutrient kg/ha*		
			N	P	K
2.	Kharif Black soils of Rajkot (Gujarat)	Groundnut	0	25	0
		Sorghum	0	30	0
		Pearl miller	0	20	0
		Cotton	0	25	0
		Sesamum	0	25	0
		Castor	0	40	0
		Green-gram	0	40	0
		Pigeon pea	0	40	0
3.	Kharif and rabi Black soils of Sholapur (Maharashtra)	Bajra	50	25	0
		Rabi Sorghum	50	0	0
		Safflower	50	25	0
		Bengal gram	15	25	0
4.	Indore(Madhya Pradesh)	Sorghum	50	25	0
		Maize	50	25	0
		Wheat	30	30	0
		Safflower	40	40	0
		Bengal gram	20	40	0
5.	Udaipur(Rajasthan)	Maize	50	30	0
		Sorghum	50	30	0
		Bengal gram	15	30	0
		Wheat	30	15	0
		Safflower	30	15	0
		Mustard	30	15	0
6.	Rabi black soils Bellary Karnataka	Rabi Sorghum	30	30	0
		Safflower	20	30	0
		Bengal gram	15	30	0
		Field bean	15	30	0
7.	Kovilpatti(Tamilnadu)	Sorghum	40	20	0
		Pearl millet	40	20	0
		Pulses	20	40	0
8.	Alluvial soils Agra(U.P)	Bajra	60	30	0
		Black&Red gram	10	40	0
		Guar	10	20	0
		Mustard	60	40	0
		Safflower	60	40	0
		Bengal gram	10	25	0
		Barley	60	0	0
		Wheat	40	0	0
		Linseed	40	20	0

Contd...

Sl No.	Region	Crop	Quantity of nutrient kg/ha		
			N	P	K
9.	Semi-arid Red soils(A.P)	Groundnut	20	40	40
		Castor	40	40	40
		Red gram	20	40	20
		Bajra	40	40	40
10.	Hyderabad(Andhra Pradesh)	Sorghum	40	30	0
		Castor	50	30	0
		Red gram	10	30	0
		Ragi&Bajra	40	30	0
11.	Bangalore (Karnataka)	Ragi	50	50	25
		Maize	75	50	25
		Groundnut	25	50	25
		Pulses	25	50	25
12.	Sub humid red soils Bhubaneshwar (Orissa)	Upland rice and medium rice	75-90	60	40
		Barley	30	20	20
13.	Ranchi (Bihar)	Rice (HYV)	60	30	0
		Rice (local)	30	20	0
		Wheat	30	20	0
		Safflower	20	0	0
		Bengal gram	0	20	6

Out of the three primary nutrients, nitrogen and phosphorus are applied in significant amount. In scanty rainfall areas, potassium seldom becomes a limiting factor jn crop yield. Its use is, therefore, avoided in most crops.

Improvement of Fertilizer Use Efficiency

The objective behind the use of fertilizer in dry land condition should be to make maximum amount of fertilizer nutrient available to the plant within the existing low levels of soil moisture. Certain points, discussed below, must be kept in mind in order to achieve this objective.

Use of Organic Matter

Organic matter constitutes a vital component of the soil. It is made up of plant residues, animal residues and dead and decayed soil organisms. To maintain a fair level of organic matter in the soil, use of farmyard manure, green manure, compost, and other organic manures is essential. The size of well decayed organic matter particles is more or less equal to the finest particles of the soil called clay. But organic matter particles play more important role than clay in retention of soil moisture. A unit volume of organic matter particles can retain 10-100 times greater amount of moisture than the same volume of clay. Use of materials supplementing to organic matter reserves of the soil is therefore, imperative in dry land soils. Besides augmenting moisture retention in the soil, organic matter helps in the development of granular soil structure which provides good aeration to the soil. Roots find congenial atmosphere for proliferation it and can extract greater amount of moisture and nutrients from the soil.

Green manuring in assured rainfall areas is very useful. Legume crops enrich soil fertility by fixing nitrogen and supplying organic matter to the soil. Growing of green gram, cowpea, black gram and soyabean with early monsoon showers for grain and by incorporating residues in the soil for August sown sorghum is a useful practice. On roadsides and field boundaries which are lying unused, leguminous shrubs and trees can be grown and the leaves and soft twigs can be cut, collected and transported to the main field. About 10 to 30 kg of fertilizer nitrogen per hectare can be saved if green manuring is done for the main crop.

Placement of Fertilizers

Placing the fertilizers in the root zone of the soil is a more efficient techniques than broadcasting of fertilizers. In the placement application, fertilizer is placed in furrows behind the desi plough at appropriate depth and at safe distance from the seeds or seedlings. By this method, contact between fertilizer and soil particles is reduced and unavailability of fertilizer nutrients due to their fixation with the soil particles can be avoided to a large extent. Fertilizer nutrients can manage to reach plant roots even if moisture in the soil is scanty.

Technique involves opening furrows towards one or both the sides of the row of crop plants, placing the fertilizer in these furrows and covering them. Furrows can be opened with the help of bullock drawn plough. Some bullock drawn machines called 'seed cum fertilizer drill' are also used for this purpose in some parts of the country. In orchard trees, fertilizer is placed in a ring which is dug around the tree at a safe distance from the tree trunk.

Spray Application

Spray application of urea and micronutrient fertilizer to supply nitrogen and micronutrients respectively, to the plants has been found to be very useful in dryland conditions. When moisture in the soil is not sufficient to transport fertilizer nutrient from the soil to the plant roots, use of, spray solution is desirable. This is an effective method of application of nitrogen through urea. Spray grade urea is marketed in our country. A 2-3% solution of urea in water can be sprayed on the crop. Two to three sprays of this kind can be given to the standing crop. If a farmer fails to apply fertilizer to his crop in time due to some circumstances such as lack of timely rainfall, unavailability of fertilizer in time and inability to farmer to buy fertilizer in time etc., application of 2-3 % urea, solution proves very beneficial.

Balanced use of Fertilizers

The singular application of either nitrogen or phosphorus or potash is not as useful as the application of their combinations. Use of phosphatic fertilizer along with nitrogenous fertilizer is found specially beneficial in dryland conditions. Phosphorus helps in elongation of root network of the plant, thereby enabling the plant roots to explore greater volume of ambient soil for nutrient and moisture. Elongated roots absorb greater amount of nutrients from the soil like nitrogen, potash, calcium etc. On the other hand, nitrogen along with other nutrients helps in development of sound foliage. The ultimate impact of the combined use of nutrients is the higher yield of crop.

Use of Phosphorus Mixed with Farmyard Manure (FYM)

Farmyard manure prepared by this method is called enriched FYM. About 750 kg of FYM is sufficient for one hectare. The whole amount of phosphatic fertilizer, calculated on the basis of recommended dose of phosphorus, for a planned crop is mixed with farmyard manure before its application to the field. Fertilizer and FYM are spread in alternate layers over one another and left in the manure pit for ripening for about a month. Pit should be covered with soil to make it airtight. Enrichment of FYM by phosphorus also increases nitrogen content of FYM. Phosphorus of FYM does not allow gaseous loss of nitrogen from FYM as it absorbs nitrogen and withholds it within the manure. Use of enriched FYM has been found very useful in dryland soils.

Use of Micronutrients and Amendments

Micronutrients are essential for the plants though the plants require them in very minute amounts. Sometime, plant growth becomes impossible due to the absence of micronutrients in the soil. Crop yield is drastically reduced due to the deficiency of micronutrients and use of fertilizers does not respond significantly. In such a situation, a representative soil sample should be taken to a local soil testing laboratory and tested there for nutrient status of the soil. Imbalance of nutrients is very likely when soil reaction is not neutral. Soil testing report is supposed to give information about soil reaction, accordingly, appropriate amount of lime in acidic soils and gypsum in alkali soils should be applied. Nutrient re- commendations based on soil test are always more desirable than blanket recommendations.

Use of bio-fertilizers

Use of bio-fertilizers is very beneficial in dry land situations. Treating the leguminous crop seeds with Rhizobium bio-fertilizers increases capacity of the crop to fix atmospheric nitrogen and dependence of plant on soil nitrogen is greatly reduced. Roots and other residues of legume crops left in the soil after harvesting, enrich the soil with nitrogen and subsequent crops get good amount of nitrogen from this soil. Mycorrhiza which is a promising biofertilizers may prove very beneficial for supporting phosphorus to crops. This biofertilizer consists of an organism called fungi. The body of micorrhizal fungi. is like a network of fine threads called hyphae. Hyphae get wrapped over plant roots and extend into the soil solution. Hyphae absorb plant nutrient from the soil, transport them to vascular bundles of the plant roots and leave them there for plant use. Application of other bulky bioertilizers. like blue green algae and azolla is also helpful in retention of moisture and increasing recovery of fertilizer used.

Alteration in Time of Fertilizer Application

Crops need fertilizer at a particular stage of growth. But a slight modification in time of fertilizer application based on moisture availability in soil, is desirable. If rainfall takes place shortly before or after the usual time of fertilizer application can be made to suit the nutrient availability. Fertilizer should be applied when there is adequate moisture in the soil. This results in greater utilization of nutrients by plants. In those dry land fields where irrigation facility is limited, calibration in timings of fertilizer application can be made accordingly.

Water Conservation Practices

It is obvious from the term dry land farming that, water deficiency is a perpetual problem and the biggest constraint to adoption of fertilizer use technology. Water conservation practices tend to encourage fertilizer application in dryland areas. All the measures should be adopted to arrest every drop of rainwater on the field itself. This will ensure greater utilization of native plant nutrients and consequently greater yield of crops.

In those dry land areas where soil and water conservation practices are adopted, fertilizer consumption is greater than that in the areas with no water conservation practices. Beneficial effect of fertilizer application coupled with moisture conservation practice. Dose of fertilizer recommended for wheat in dryland condition was applied; moisture level was measured as thickness of water layer in millimeter, in the top 90 cm thick vertical layer of soil.

Table: *Effect of moisture conservation practices on wheat yield*

Sl.No	*Treatment*	*Moisture level in soil profile (mm/90cm)*	*Yield as average of two years (kg/ha)*
1	Farmers practice	147	1500
2	Minor leveling& bundling	168	1780

Graded bunding with water ways should be preferred ~ for all black and heavy textured soils like clay, clay loam and silty clay. Graded bunding is also suitable for all areas receiving more than 800 mm of rainfall and soils having impervious layers within a depth of 22.5 cm irrespective of soil type. Contour bunding is recommended for areas receiving less than 800 mm rainfall and having light to medium texture soils with good permeability. For the slopes, terracing is recommended. Whereas in the cultivated land deep ploughing is an easy method of water conservation.

Other Farming Practices

Other farming practices should be based on two basic necessities of dryland farming namely moisture and nutrient supply. In kharif season nitrogen should be given in two or three split doses in order to avoid leaching loss of nitrogen. However, for rabi season crops, nitrogen can be placed along with phosphorus and potash before sowing. Proper sequence of crops (rotation) help in balanced use of nutrients in the soil. Crop rotation must include at least one leguminous crop.

Certified quality seeds should be preferred to traditional ones. It is often observed that the high yielding varieties of sorghum, pearl millet, castor, and beans, adapted to that particular region, have deeper and more extensive root systems than the local varieties. Hence, they are able to withstand moisture stress and absorb greater amount of nutrient from the soil.

Some farming practices specific to dryland conditions like manipulation in sowing time, following cropping pattern and cropping intensity recommended for that particular area, proper insect pest and weed control etc. must accompany the use of fertilizers.

Mulching

The practice of mulching involves placing materials such as straw, plant residues leaves, loose soil or plastic film on the soil surface to reduce loss of moisture by evaporation from the soil. Mulching reduces soil erosion and also protects the plant roots against wide fluctuation in soil temperature.

Mulching of rainfed crops helps retain soil moisture and increase fertilizer use efficiency. Yield of wheat often increases if mulching is done in its preceding crop of maize. In wheat, following mulched maize, the yields with 0 and 40 kg N/ha were comparable to those with 40 and 80 kg N/ha without mulching. Thus, it saved 40 kg N/ha to the succeeding crops of wheat.

Conclusion

To meet the food grain requirement of 200 million additional population by 2000 A.D., we have to produce 70 million tonnes of food grains over and above what we produce today (target of 2000 A.D. is 240 million tonnes).

More than 80 % of the additional requirement of food has to be met from the dryland area in nineties. Proper application of the appropriate technology, of proven (benefit, must receive attention. The package of practices of dryland farming evolved by the various research institutions located in various parts of India have made the fertilizer use technology free from risk, cost effective and ecologically sound. Farmers in dryland areas are very receptive to fertilizers.

Yet, a large number of them stay away from fertilizers as the use of fertilizers involves a risk because 1 of uncertainty of monsoon in India and a very short 1 span (only 2-3 months) of rainfall.

The mean length of growing season varies from 18 weeks on shallow red soils, 22 weeks on deep red soils to 22 weeks on deep black soils.

The calendar of operations should be so framed as to take the fullest advantage of growing season. Investigations have shown that in assured rainfall zones of dry land regions, investing one rupee in fertilizer gives return of 2 to 15 rupees in monsoon season.

In post monsoon season the return has been worked out to be 2 to 12 rupees for each rupee invested in fertilizer, in assured rainfall zone.

To realize this much profit, fertilizer application must be accompanied by other improved practices of dry land farming.

Improved Dryland Technologies and Weed Management

Intercropping: Mixed cropping is traditionally adopted in rained areas for risk distribution during adverse weather conditions there by achieving yield stability.

Now the mixed cropping has changed to intercropping due to substantial research efforts. The intercropping has proved substantial yield advantage over sole crop. It also gives greater stability of yields over different seasons.

The intercropping is more popular among the small farmers to meet their need for different crops. The intercropping is more advantageous during Kharif than rabi season.

Points to be considered while selecting crops for intercropping:

i) Both the crops should have differential growth habit
ii) They should have different crop growth period.
iii) There should be more competition tree period between these two crops.
iv) They should have differential requirements interspect at nutrients and soil moisture.
v) They should have different feeding zones.
vi) Mostly there should be a combination of cereal and legume.

Eg .Bajra + Tur - 2:1 proportion

Cotton + green gram - 2: 1 proportion

Sorghum + green gram 2: 1 proportion

Sunflower + Tur - 2: 1 proportion

Sorghum + Tur 2: 1 proportion

Fertilizer Use in Dryland

Fertilizer use is an reertant factor for increasing the yield of dryland crops. The soils of drylancare low in N low to medium in P and rich in available K.

The research breeds out in fertilizer use of various crops have suggested ad definite resense to fertilize, application. The fertilizer application not only increase are vield of dry land crops but also enhance the growth and maturity of the nops.

Based on the results of various trials conducted at Solapur the nacwing economic fertilizer doses are recommended for dryland crops.

Crops	*Recommended fertilize does (Kg/ha)*		
	N	*P205*	*K2S*
A) Scarcity zone (Solapur)			
Kharif			
1. Bajra	50	25	-
2. Tur			
3. Gr. Nut	12.5	25	-
4. Horse gram			
5. Kidney bean			
6. Sunflower	50	25	-
7. Castor	25	12.5	
Rabi			
1. Sorghum	50	-	-
2. Safflower	50	25	-
3. Gram	12.5	25	-
B) Assured rainfall zone (Akola)			
1. Jowar	80	40	-
2. Rala	60	30	-
3. Cotton	80	40	-
4. Tur, Mung, Urid, Gr. Nut	20	40	-

The time and method of application of fertilizer are also imcutant because of moisture limitation of dryland. The fertilizer should be applied at the time of sowing by placement method. For this purpose, Solapur centre has developed two bowl ferti seed drill which now become more sepular among the farmers. The response top application to deep black soils is limited the to high p fixation due to calcium content. For Rabi crops advance placement of p by 10 - 20 days before actual sowing is usual practice. The soils of dryland are medium to rich in available K and hence no sponse to K application.

Under light textured soils boron application

@ 5 kg Borax x / ha at in ernate years to groundnut is recommended for dry lands of Maharashtra.

Under intercropping systems, there is no need to apply additional utilizes to intercrop of legume under cereal + legume combination.

Under sequence cropping under dry lands, when cereal is grown after gumes like green gram, black gram, the N application may be reduced by 5 to 30 kg N / ha to cereal or oil seed crops like sorghum sunflower of safflower.

Weed Management: It is estimated that weed cause about 37 to 79% pass to the crop production in dryland agriculture where moisture is the most limiting factor. The research findings on weed - crop competition in Dryland indicated that in crops with 80 to 120 days duration first 20 to 30 lays period is more sensitive to competition from weeds. Checking weed growth during this period helps in to minimize the loss in crop production. The results with perarlmillet at Solapur confirm the above observations.

Weed control through chemicals: Chemicals are found to be as effective as mechanical measures in controlling the weeds in dryland agriculture. Use of Atrazine @ 0.5 to 1.0 kg a.i. / ha as pre-emergence for cereal crop is effective. The weeds like striga can be controlled effectively by 2, 4 - D from mechanical measures.

Integrated weed control: A combination of mechanical and chemical weed control methods are most effective. Studies carried out at Solapur on methods are most effective. Studies carried out at Solapur on Kharif groundnut showed that combination of pre - emergence spray of Propanil @ 3 kg/ha + weeding at 30 days after sowing increased yield of groundnut by 100 per cent. Use of Basaline ! 1 kg/ha + one hand weeding at 30 DAS has also found effective at other locations.

Off Season Tillage for Weed Control: Off season tillage would control the weeds reaching maturity after the harvest of first crop and checking multiplication. It also helps in timely sowing of crop with first shower of rain.

Rainfed Agriculture

The term Rainfed agriculture is used to describe farming practises that rely on rainfall for water. It provides much of the food consumed by poor communities in developing countries. For example, rainfed agriculture accounts for more than 95% of farmed land in sub-Saharan Africa, 90% in Latin America, 75% in the Near East and North Africa; 65% in East Asia and 60% in South Asia.

Levels of productivity, particularly in parts of sub-Saharan Africa and South Asia, are low duc to degraded soils, high levels of evaporation, droughts, floods and a general lack of effective water management. A major study into water use by agriculture, known as the Comprehensive Assessment of Water Management in Agriculture, coordinated by the International Water Management Institute, noted a close correlation between hunger, poverty and water. However, it concluded that there was much opportunity to raise productivity from rainfed farming.

The authors considered that managing rainwater and soil moisture more effectively, and using supplemental and small-scale irrigation, held the key to helping the greatest number of poor people. It called for a new era of water investments and policies for upgrading rainfed agriculture that would go beyond controlling field-level soil and water to bring new freshwater sources through better local management of rainfall and runoff.

The importance of rainfed agriculture varies regionally but produces most food for poor communities in developing countries. In subSaharan Africa more than 95% of the farmed land is rainfed, while the corresponding ûgure for Latin America is almost 90%, for South Asia about 60%, for East Asia 65% and for the Near East and North Africa 75% (FAOSTAT, 2005). Most countries in the world depend primarily on rainfed agriculture for their grain food. Despite large strides made in improving productivity and environmental conditions in many developing countries, a great number of poor families in Africa and Asia still face poverty, hunger, food insecurity and malnutrition where rainfed agriculture is the main agricultural activity. These problems are exacerbated by adverse biophysical growing conditions and the poor socioeconomic infrastructure in many areas in the semi-arid tropics (SAT). The SAT is the home to 38% of the developing countries' poor, 75% of whom live in rural areas. Over 45% of the world's hungry and more than 70% of its malnourished children live in the SAT

There is a correlation between poverty, hunger and water stress (Falkenmark, 1986). The UN Millennium Development Project has identiûed the 'hot spot' countries in the world suffering from the largest prevalence of malnourishment. These countries coincide closely with those located in the semi-arid and dry subhumid hydroclimates in the world, i.e. savannahs and steppe ecosystems, where rainfed agriculture is the dominating source of food and where water constitutes a key limiting factor to crop growth (SEI, 2005). Of the 850 million undernourished people in the world, essentially all live in poor, developing countries, which predominantly are located in tropical regions (UNSTAT, 2005).

Since the late 1960s, agricultural land use has expanded by 20–25%, which has contributed to approximately 30% of the overall grain production growth during the period (FAO, 2002; Ramankutty et al., 2002). The remaining yield outputs originated from intensiûcation through yield increases per unit land area. However, the regional variation is large, as is the difference between irrigated and rainfed agriculture. In developing countries rainfed grain yields are on average 1.5 t/ha, compared with 3.1 t/ha for irrigated yields (Rosegrant et al., 2002), and increase in production from rainfed agriculture has mainly originated from land expansion. Trends are clearly different for different regions. With 99% rainfed production of main cereals such as maize, millet and sorghum, the cultivated cereal area in sub-Saharan Africa has doubled since 1960 while the yield per unit of land has been nearly stagnant for these staple crops (FAOSTAT, 2005).

In South Asia, there has been a major shift away from more drought-tolerant, low-yielding crops such as sorghum and millet, while wheat and maize hasapproximately doubled in area since 1961 (FAOSTAT, 2005). During the same period, the yield per unit of land for maize and wheat has more than doubled. For predominantly rainfed systems, maize crops per unit of land have nearly tripled and wheat more than doubled during the same time period.

Rainfed maize yield differs substantially between regions. In Latin America (including the Caribbean) it exceeds 3 t/ha, while in South Asia it is around 2 t/ha and in subSaharan Africa it only just exceeds 1 t/ha. This can be compared with maize yields in the USA or southern Europe, which normally amount to approximately 7–10 t/ha (most maize in these regions is irrigated).

The average regional yield per unit of land for wheat in Latin America (including the Caribbean) and South Asia is similar to the average yield output of 2.5–2.7 t/ha in North America. In comparison, wheat yield in Western Europe is approximately twice as large (5 t/ ha), while in sub-Saharan Africa it remains below 2 t/ha. In view of the historic regional difference in development of yields, there appears to exist a signiûcant potential for raised yields in rainfed agriculture, particularly in sub-Saharan Africa and South Asia.

Rural development through sustainable management of land and water resources gives a plausible solution for alleviating rural poverty and improving the livelihoods of the rural poor. In an effective convergence mode for improving the rural livelihoods in the target districts, with watersheds as the operational units, a holistic integrated systems approach by drawing attention to the past experiences, existing opportunities and skills, and supported partnerships can enable change and improve the livelihoods of the rural poor. The well-being of the rural poor depends on fostering their fair and equitable access to productive resources.

The rationale behind convergence through watersheds has been that these watersheds help in 'cross-learning' and drawing on a wide range of experiences from different sectors. A significant conclusion is that there should be a balance between attending to needs and priorities of rural livelihoods and enhancing positive directions of change by building effective and sustainable partnerships. Based on the experience and performance of the existing integrated community watersheds in different socioeconomic environments, appropriate exit strategies, which include proper sequencing of interventions, building up of financial, technical and organizational capacity of local communities to internalize and sustain interventions, and the requirement for any minimal external technical and organizational support needs to be identified.

Dry Land Farming With Special Reference to Telangana

Dry land farming is an agricultural technique for non-irrigated cultivation of land. It may be defined as: "a practice of growing profitable crops without irrigation in areas which receive an annual rainfall of 500 mm or even less". India has about 108 million hectares of rain fed area which constitutes nearly 75% of the total 143 million hectares of arable land. In India, dry land agriculture accounts for nearly two-thirds of total cropped area and generates nearly half of the total value of agricultural output. Rain fed agriculture encounters several constraints on account of climatic, edaphic, and social factors. Out of the 97 million farm holdings in India, about 76% come under marginal and small categories. The productivity levels of these areas have remained lower across years because of frequent droughts occurring due to high variability in the quantum and distribution of rainfall, poor soil, low fertilizer use, imbalanced fertilization, small farm size and poor mechanization, poor socio-economic conditions and low risk-bearing capacity, low credit availability and infrastructure constraints. Consequently, farmers are distracted from agriculture and tend to migrate to cities to look for alternative jobs. Hence, there is a great need to increase the productivity of rain fed crops and overall net returns to keep the farmers in agriculture. A paradigm shift in rain fed agriculture can be expected through technological thrusts and policy changes.

India has about 47 million hectares of dry lands out of 108 million hectares of total rain fed area. Dry lands contribute 42% of the total food grain production of the country. These areas produce 75% of pulses and more than 90% of sorghum, millet, groundnut and pulses from arid and semi-arid regions. Thus, dry lands and rainfed farming will continue to play a dominant role in agricultural production.

Dry lands, besides being water deficient, are characterized by high evaporation rates, exceptionally high day temperature during summer, low humidity and high run off and soil erosion. The soils of such areas are often found to be saline and low in fertility. As water is the most important factor of crop production, inadequacy and uncertainty of rainfall often cause partial or complete failure of the crops which leads to period of scarcities and famines. Thus the life of both human being and cattle in such areas becomes difficult and insecure.

Despite all improvements in agriculture, we have not yet been able to evolve appropriate practices for our dry land areas. The income of farmers of dry land regions is low. We continue to stress on intensive agriculture on irrigated land but we cannot afford to be complacent with our dry lands. We cannot achieve stability in food production with unstabilized dry land agriculture. Therefore, we are required to adopt improved technology especially developed for dry land agriculture.

The strategies that need to be emphasized:

(1) Land care and soil-quality improvement through conservation agricultural practices, balanced fertilization, harnessing the potential of bio-fertilizers and microorganisms, and carbon sequestration;

(ii) Efficient crops, cropping systems, and best plant types;

(iii) Management of land and water on watershed basis;

(iv) Adoption of a farming-systems approach by diversifying enterprises with high-income modules;

(v) Mechanization for timely agricultural operations and precision agricultural approach;

(vi) Post-harvest, cold-storage, value-addition modules;

(vii) Assured employment and wage system;

(viii) Organic farming;

(ix) Rehabilitation of rain fed wastelands;

(x) Policy changes and other support system; and

(xi) Human-resource development, training and consultancy.

Characteristics of Dry-land Agriculture:

Dry land areas may be characterized by the following features:

1. Uncertain, ill-.distributed and limited annual rainfall;
2. Occurrence of extensive climatic hazards like drought, flood etc;
3. Undulating soil surface;
4. Occurrence of extensive and large holdings;
5. Practice of extensive agriculture i.e. prevalence of mono-cropping etc;
6. Relatively large size of fields;

7. Similarity in types of crops raised by almost all the farmers of a particular region;
8. Very low crop yield;
9. Poor market facility for the produce;
10. Poor economy of the farmers; and
11. Poor health of cattle as well as farmers.

Problems of Dry Land Farming in India

The major problem which the farmers have to face very often is to keep the crop plants alive and to get some economic returns from the crop production. But this single problem is influenced by several factors, they are:

Moisture Stress and Uncertain Rainfall: The rains are very erratic, uncertain and unevenly distributed. Therefore, the agriculture in these areas has become a sort of gamble with the nature and very often the crops have to face climatic hazards. The farmers also take up farming halfheartedly as they are not sure of being able to harvest the crops. Thus, water scarcity becomes a serious bottleneck in dry land agriculture.

Effective Storage of Rain Water: According to characteristics of dry farming, either there will be no rain at all or there will be torrential rain with very high intensity. Thus, in the former case the crops will have to suffer a severe drought and in the latter case they suffer either flood or water logging and they will be spoilt In case of very heavy downpour, the excess water gets lost as run-off which goes to the ponds and ditches etc. This water could be stored for providing life saving or protective irrigation to the crops grown in dry land areas. The loss of water takes place in several ways namely run-off, evaporation, uptake through weeds etc.

Marketing Problem

In dry farming all the farmers grow similar crops which are drought resistant. These crops mature at the same time and the growers like to dispose off their products soon after the harvest. This results in a glut of products in the market and the situation is badly exploited by the grain traders and middlemen. Therefore, marketing becomes a serious problem in dry farming areas.

Unbalance Their Economic Position

Only drought resistant crops namely oilseeds, pulses and coarse grains like jowar, bajra, millets etc. can be grown in dry land areas. Thus, the farmers have to purchase other food grains and household commodities that unbalance their economic position.

Careful and Judicious Manorial Scheduling

In case of irrigated farming the farmers are at a liberty to apply manures and fertilizers according to their availability and facility but in case of dry farming they have to be very careful in fertilizer application. Due to lack of available moisture, broadcasting or top dressing becomes wasteful and meaningless. These can be applied by only deep placement and foliar spray for an improved crop production.

Utilization of Preserved Moisture

Judicious and purposeful utilization of preserved moisture water depends upon soil type, plant type and other factors. The amount of available water to the plants depends upon the depth of plant roots, their proliferation and density. In case of limited moisture condition, the yield directly depends upon the rooting depth. The rooting depth can be desirably increased by mechanical manipulation of the soil. If the planting is very dense and all the plants have same kind of rooting then there will be a tough competition among roots for moisture and scarce moisture condition will result in the wilting of plants. Therefore, utilization of preserved moisture is an art in dry farming. The water collected in ponds or brooks may be used to give protective or life saving irrigation. The widely spaced crops can be intercropped with oilseeds or pulses for increasing the productivity of the land per unit area and per unit time. Therefore, the water collected during the rainy season need special technique and skill for its efficient utilization.

Quality or the Produce

The quality of the produce from dry farming areas is often found to be inferior as the grains are not fully developed or they are not filled properly; often mixed with other crop seeds owing to mixed .cropping system prevalent in these areas and the fodder become more fibrous. All these factors reduce the market value of produce and the farmers do not get the profit of their labour and investment.

Special Reference to Telangana

Telangana, the semi-arid land of India, is experiencing drought often pushing large numbers of people to the margins of living. Drought visits south Telangana "once in two and half years". The rainfall of about 70 cm and less in southern Telangana hardly justifies the fact that the region should languish under semi-arid conditions. In fact, the region forms part of the catchment of the perennial rivers Krishna and Godavari. The irrigation policy initiative over the years continuously favoured the Delta region leaving a large number of people at the mercy of degraded nature and sub-human living. Thus "Telangana backwardness has essentially political roots: with better administration the considerable water resources could have been more fully tapped for irrigation. Telangana is still mainly a dry farming area, the reason for this in Telangana is long term failure to harness the potentialities of the area."

Analysis: As major irrigation facilities are not sufficiently available and short fall of normal rain, when compared to Costal Andhra region, Telangana is mostly dependent on dry land farming. In Telangana, in view of the non-availability of water in wells and tubewells due to depletion of water table and drought farmers are now forced to keep major part of their dry lands as current follow lands. As a result, the area under follow land is increasing year after year. There is a shift in cropping pattern and at the same time, the total area under jowar, bazra, caster and cereals decreased significantly, while the cultivable area under rice, maize, groundnut, oilseeds, cotton, pulses increased proportionately. The yield and production of these crops also increased. It is a definite change in favour of commercial crops.

Due to recurring drought conditions, most of the borrowers in rural areas of Telangana could not repay the loans borrowed earlier. In view of this, financial institutions kept those villages as de-faulted borrowers, included in the black list closing their chance of borrowing again. This has become a stumbling block to majority of the rural households in all the regions in the state, particularly in Telangana. Consequently, the dependency on money lenders and private financiers is again on the increase lending to increase in the cost of production, unremunerative cultivation and increasing indebtedness.

The State needs to give priority for agriculture particularly, in the field of irrigation sector and cheap and assured credit facility. The focus should be on dry land farming, extension services and provision of quality seeds and fertilizers and timely assistance. In recent years the plan allocations to the priority sectors such as agriculture, irrigation have been declining from plan to plan. Irrigation sector is neglected.

Many of the proposed projects in Telangana region could not be undertaken. While total canal irrigation through canals remained stagnant, tank irrigation declined during the last two decades. Similarly, cultivation under dugwells and borewells has increased significantly leading to power problems, and depleting water table below 600 feet in certain areas, and gradual withdrawal of subsidies to agricultural sector also increased cost of cultivation unremunerative cultivation. This has led to unrest among the farmers resulting suicide deaths especially in Telangana region.

Telangana projects have been allocated 266.83 TMC (Thousand Million Cubic Feet. One TMC ft is equivalent to about 28.317 million cubic metres or 22 956.8 acre feet) of water against a due share of 552 TMC. Mahboob Nagar known for its very high levels of distress migration and perennial drought, should have got 187 TMC of water but have received nothing till now. Costal Andhra receives several times more than its due allocation of 99 TMC. Farming has become risky in Telangana, as indicated in the large number of suicides by farmers. Telangana accounts for as many as two- thirds (66%) of the total number of suicides reported in the state between 1998 and 2006. Though recent data shows that Telangana has been allocated a higher share in expenditure on irrigation (55%) than its share of population (41%) however, compared to costal Andhra, the unit cost of irrigation is much higher in Telangana (as it is situated in Deccan plateau) as lifting of water requires huge investments in pumping machinery and power.

As a matter of fact, majority of the households in villages are considered to be labourers. Real development of villages can only be achieved if the labour households' employment, wages and incomes are improved. It is observed that employment, wages and other living conditions of labour households are further deteriorated in recent times. Non-agricultural employment is found to be significant in those areas where canal irrigation is provided. With the development of agriculture, non-agricultural employment was also generated in the command areas. Due to backward agriculture and frequent droughts in most of the mandals of Telangana, labour households find it difficult to get employment during lean seasons and prefer to migrate to far and near places. The process of migration has accelerated in recent years. Due to lopsided developmental strategies pursued from time to time, balanced development of the state has become a casualty and regional imbalances went on widening. These imbalances have become stumbling blocks for the emotional integration of the people of all the three regions of the state. Further, the process of implementation of economic reforms including privatisation is taking place at an accelerated pace in the state.

Impact of Liberalization, Privatization and Globalization

The policies of liberaliation, privatisation and globalization have been displacing the masses from their opportunities. The benefits and subsidies meant for weaker sections are reduced year after year even these meant for the Scheduled Castes and Scheduled Tribes. This is not followed by a corresponding support in alternative occupations or opportunities. With introduction of labour saving technology in the field of construction of roads and buildings, wage labourers have been badly effected in the state. In view of the lopsided pattern of development, the state has been witnessing agitations, movements, rural unrest, farmer suicides and hunger deaths in recent years. The village economy is facing economic and social crisis. Agriculture is unable to absorb the over increasing working population. Further, the cost of production per unit of agricultural output in Andhra Pradesh now is higher compared to major agricultural States in India. The area under canal irrigation system declined due to deceleration in public investment in Telangana. In nutshell, the agricultural sector is neglected by the government. There is a need to review this policy.

Small and marginal farmers have been worst effected. Majority of the small and marginal farmers still depend in informal or non-institutional sources of credit, particularly, money lenders, private financiers at higher rates of interest, consequently, high cost of production and indebtedness.

Liberalisation and Privatisaion process was initiated in the state with firm determination during 1996-'97. But it's impact is not well received by all section of the people. Agricultural growth rates have gone down drastically. Employment situation in rural areas were not improved rather deteriorated. Whatever the employment opportunities have been created so far, they are largely low paid and casual in nature and insecure. Nonagricultural employment could not be generated to the levels of expectations. Villages have become markets for products of multinational, and big industries. Whatever the industries, or small scale industrial units were available earlier, they are unable to compete with global products either in quality or prices. Backward area like Telangana (except Hyderabad) could not attract either domestic or foreign direct investment.

There is an exodus of young persons from villages of backward and drought effected districts to towns and cities in search of livelihood. People from Mahaboobnagar, Nalgonda, Karimnagar, Warangal and Medak are migrating to gulf countries and Hyderabad to get some livelihood or other. Only old age people keep staying in rural areas. The Information and Technology could provide jobs to a few thousands of educated young people.

The Self-help Groups for women could not provide work as expected. This programme could enlighten rural women groups in political and social aspects. These groups could mobilise savings out of their hard earned income besides State/ Central assistance. As far as employment and income generation activity of this programme is concerned, very little is achieved. Whatever the products are produced by these groups, they are decorative and artistic, unable to compete with global multinational products. Mostly they are neither mass consumption oriented nor essentials. Hence, they suffer from lack of demand. The scheme has become political wing of ruling party for vote bank.

Suggestions:

(1) Agriculture may be given top priority along with infrastructure development in backward regions. Constructions of Ichampally irrigation project across river Godavari will benefit north Telangana. Similarly, through proper allocation and utilisation of Krishna river water will also benefit Nalgonda, Mahabubnagar districts in South Telangana.

(2) Distribution of cultivable public lands surplus lands and cultivable waste lands among the rural poor provides some solution to the agricultural labourers.

(3) There is an urgent need to change the cropping pattern in drought prone areas of the regions to prevent further downslide of under ground water table.

(4) It is also necessary to identify backward districts and specific area programmes may be initiated through state and central grants.

(5) Rural and agricultural credit facilities have to be adequately provided to all the needy households keeping in view the growing dependency of farmers and rural artisans on money lenders and private financiers.

(6) Both central financial transfers and use of policy instruments will be useful to attract private investment to the backward regions; and

(7) Enhanced allocations for social development such as education, health, nutrition, empowerment of poor. Further democratization of rural institutions etc., will improve education, skills and entrepreneurial abilities of people in backward areas.

Conclusion: Even after utilizing all the available water resources, about 50% of our cultivable area will still depend on rains. Therefore, our agricultural scientists, policy formulators and farmers should appropriately realize the magnitude of role that. rainfed agriculture or dryland farming can play. They should thoroughly examine the problems of dry land agriculture from different viewpoints and evolve appropriate technologies, crop varieties, etc. for these areas to better the economic position of the farmers. Dry farming areas, therefore, need a much closer attention for achieving food security in India.

Problems of Crop Production in Dryland

Dry farming crops are characterized by very low and highly variable and uncertain yields. Crop failures are quite common. These are mainly due to the following causes.

Inadequate and Uneven Distribution of Rainfall

In general, the rainfall is low and highly variable which results in uncertain crop yields. Besides its uncertainty, the distribution of rainfall during the crop period is uneven, receiving high amount of rain, when it is not needed and lack of it when crop needs it.

Late Onset and Early Cessation of Rains

Due to late onset of monsoon, the sowing of crop are delayed resulting in poor yields. Sometimes the rain may cease very early in the season exposing the crop to drought during flowering and maturity stages which reduces the crop yields considerably

Prolonged Dry Spells During the Crop Period

Long breaks in the rainy season is an important feature of Indian monsoon. These intervening dry spells when prolonged during crop period reduces crop growth and yield and when unduly prolonged crops fail.

Low Moisture Retention Capacity

The crops raised on red soils, and coarse textured soil suffer due to lack of moisture whenever prolonged dry spells occur due to their low moisture holding capacity. Loss of rain occurs as runoff due to undulating and sloppy soils.

Low Fertility of Soils

Drylands are not only thirsty, but also hungry too. Soil fertility has to be increased, but there is limited scope for extensive use of chemical fertilizers due to lack of adequate soil moisture.

Agrodiversity as a Means of Sustaining Small Scale Dryland Farming Systems in Tanzania

The Issues

Dryland farming systems in Africa are often characterised as being extremely degraded, vulnerable to external forces, and low in productive output. Available fossil evidence, for example, suggests that the open savannas of East Africa have the longest history of human habitation since Pliocene times. As a result, Adams (1996, p.208) observes that, "the destructive impact of long human occupation of the ecosystem has been profound. ..... Man's tendency to overexploit the basis of his subsistence is endemic." Such stereotypes of land use abound. The story of man's mistreatment of his environment has been recounted many times and continues to be told. Lord Hailey (1938) described human-induced soil erosion as "the scourge of Africa", while the FAO (1990, p.6) bemoans that, "Africa's lands are under attack." This paper targets a very different tendency – human's propensity to conserve. Some land uses, developed over centuries of pressures and difficulties, display a resilience and productivity that is truly remarkable. Both professionals and policy makers may draw some comfort that stereotypes of environmental crisis are not applicable throughout dryland Africa.

Nevertheless, environmental change is occurring. Global biodiversity loss in areas of land use is a well-attested phenomenon. Ecologists, in particular, are alarmed at how natural biological diversity is being replaced by relative biological uniformity, especially under the pressure of population growth (Cincotta and Engelman, 2000). As natural habitats decline, greater proportions of species living within those habitats become extinct. Species-area curves suggest that at about 10 percent of land area devoted to protection, only 45 to 70 percent of species remain, and as habitat declines further, extinctions accelerate dramatically (Pimm *et al*, 1995). Only Costa Rica comes anywhere near the conservationist's goal of maintaining at least 10 percent of land areas under natural habitat. However, there is good evidence that natural biological diversity may be giving way to another diversity, equally valuable and of greater immediate significance to society, which in this paper is called 'agrodiversity'. It embodies cultural and spiritual dimensions of biodiversity (Posey, 1999), as well as practical and economic values of gaining a sustainable rural livelihood for poor people (Altieri, 1999). Without in any way denying the need to conserve natural biological diversity upon which ecologists argue many of the world's life-support functions are based, this paper dwells on the benefits of 'agrodiversity' and empirical examples from Tanzania on its value to local society and, indeed, to the whole conservation agenda.

Introducing Agrodiversity

Awareness of global biodiversity loss has resulted in a search for where the battle-lines for its protection should best be drawn. Typically, these lines have been fortified around the various types of protected areas and controlled management of habitats. Increasingly, this 'fortress conservation' has become unsustainable in the face of social change and population pressures (Ghimire and Pimbert (1997).

Additionally, although evidence for the relative numbers of species is elusive, there is likely to be far more biodiversity in areas of land use than in all protected areas together, principally because areas that are used are not only far greater than those protected but also these areas are the more fertile and naturally biodiverse. Consequently, it makes sense for the global biodiversity agenda to look to:

- natural biodiversity managed within farming systems, including domestication of wild species and protection of rare species;
- the farming systems themselves, including the way that biodiversity is managed for the welfare of land users such as techniques and practices of cropping, agroforestry and rangeland management;
- how biodiversity itself helps vulnerable farming groups cope with social, economic, political and demographic pressures as well as variable and marginal physical environments.

While the term 'agricultural biodiversity' or 'agro-biodiversity' has tended to be used to signify the variety of plants, species and varieties on the lands of farmers and their relation to welfare (Thrupp, 1998), 'agrodiversity' is much more broadly defined. It encompasses "the many ways in which farmers use the natural diversity of the environment for production, including .... their choice of crops....[and] their management of land, water, and biota" (Brookfield and Padoch, 1994, p.9). It goes beyond the concept of species and genetic diversity of plants and animals to incorporate other aspects of the farming system that relate to obtaining sustainable livelihoods.

Because small-scale farmers in the tropics usually have to rely on the intrinsic quality of their natural resources, including biodiversity and soil quality, and because they have limited resources to invest in external inputs, 'agrodiversity' is a crucial underpinning to their lives.

This paper, therefore, illustrates some aspects of the importance of agrodiversity to the ways in which small-scale farmers in a dryland part of Tanzania cope with environmental and social pressures. It takes up Paul Richards' (1985, p.160) challenge that, "the evidence .... of innovativeness in the peasant food crop sector is strong, but further work is needed on ...how....this pool of skill and initiative might be harnessed to national development objectives."

The policy implications of agrodiversity will be developed in the last section of this paper. First, however, we describe the approach of the GEF-funded project, *People, Land Management and Environmental Change*, which is compiling a database of agrodiversity in order to demonstrate its value and potential.

PLEC Approach

The aim of the People, Land Management and Environmental Change (PLEC) project is to develop sustainable and participatory approaches to conservation within small farmers' agricultural systems, and in participation with farmers.

The specific objectives of PLEC are:

a) to establish historical and baseline comparative information on agrodiversity, including biodiversity, at the landscape level;
b) to develop participatory and sustainable models of biodiversity management based on farmers' technologies and knowledge within agricultural systems at the community and small-area levels;
c) to recommend approaches and policies for sustainable agrodiversity management to key government decision makers, farmers, and field practitioners; and
d) to establish national and regional networks for capacity strengthening within participating institutions, and to carry forward the aims of PLEC.

The core of PLEC's work is in its 'demonstration site' villages. Here, PLEC becomes the farmers' own enterprise, and scientists are the facilitators, not the instructors. The scientists identify and demonstrate farmers' practices that are environmentally, socially and financially sustainable, and which sustain biodiversity. They help farmers in achieving their own conservationist goals. Collaborating farmers manage varied biophysical conditions, growing a range of crops and using biodiversity with discretion. The PLEC approach differs from mainstream agricultural research at experiment stations under controlled conditions. By integrating locally developed knowledge of soil, climate, and other physical factors with scientific assessments of their quality in relation to crop production, a set of sustainable agricultural technologies can be devised so that agricultural diversity is maintained. The participatory process will eventually enhance farmers' and local communities' ability to adapt to environmental, social and economic change.

The PLEC project, through demonstration sites and farmer-to-farmer extension, seeks to support existing diversity and disseminate it to other farmers whose positions might be improved as a result of new practices or techniques. PLEC relies on interaction between farmers and between farmers and scientists as a mechanism for identifying and maximising agrodiversity. The Tanzanian element of the East African PLEC cluster focuses on two contrasting demonstration sites in Arusha region in Northern Tanzania. The characteristics of each are summarised in the following table.

Table: *Characteristics of PLEC sites in Northern Tanzania.*

Characteristic	*Ngiresi/Olgilai*	*Kiserian*
Average altitude	1,900m asl	1,200m asl
Temperature range	12-30° C	12-30° C
Mean annual rainfall	2,000mm	500mm
Farming system	Mixed cropping with zero grazing	Agropastoral
Soil type	Andosols	Cambisols
Village population (1988)	2,158	3,330

In Tanzania the PLEC scientists initially met with farmers to identify environmental and social constraints and to see how coping strategies were related to those constraints. Following on from this, 'expert' farmers were identified and, using PRA techniques and training sessions, linkages were established with other farmers. The focus has been on identifying ways in which farmers have adapted their practices to, and have made use of, the environment in which they farm while at the same time conserving or enhancing agrodiversity, especially biodiversity.

Diversity in Crops and Cropping Systems

Planting a mix of crops is recognised as a way of spreading risk on a farm. Within crop groups, different varieties are planted to match the particular stresses of the local environment. Farmers seek to plant a combination of varieties that will ensure at least some yield despite extremes of climate, pests, disease, labour shortage or other constraints.

In a similar way, farmers select the crop types that they plant to enhance their food security. For example, in Ngiresi sweet potatoes are planted on steep slopes not only to provide cover and thereby reduce erosion, but also because they mature in the 'hungry period' before maize is ready for harvest. In addition, sweet potatoes will produce even in a drought year.

Farmers, therefore, select crops and varieties using different criteria – some strains will be selected because they are high yielding in optimum conditions, others because they are tolerant to drought, others because they are resistant to storage pests, have a high market price, good taste or are easily processed. Each selection involves an assessment of the potential risks and rewards of planting a particular crop or variety. Such decisions are influenced not only by the physical characteristics of the environment, but also by socio-economic factors such as available labour and proximity to markets.

Farmers view the selection of crop varieties as a continuous process. Some varieties that are tried in the field become part of the farmer's own landrace, whereas others, whose characteristics prove to be less suited to the local environment, quickly disappear from the field. Crops and varieties that continue are those identified as being a best match for the field conditions, the wider environment and the farmer's own situation.

In the Tanzanian PLEC sites maize, beans and bananas are very widely grown. Within these crop types a wide spectrum of varieties is planted. The following table lists some of the varieties identified on farms. This table clearly illustrates that, while research stations may breed for high yields, drought tolerance and pest resistance, farmers consider a wide range of other characteristics before selecting which seeds to plant.

Issues such as intercropping compatibility and labour availability are also relevant to farmer decision-making. So for example, trailing beans are less popular than the bush varieties because they get entangled with the maize intercrop. More labour is required to disentangle them and beans are lost when pods burst during this process. Most farmers plant up to five different varieties of beans and bananas on their land.

This reflects not only risk-spreading decisions, but also the matching of particular varieties to scattered plots having different biophysical characteristics. Traditionally, maize and beans have been intercropped in the two sites. No standard proportions are used, rather the exact combinations grown depend on assessments by individual farmers of the specific plot, previous experience on that plot, market conditions, location and topography of the field etc. Typically, in Kiserian, a higher proportion of beans are planted (2 rows of beans for every row of maize). This reflects not only the fact that beans are a highly valued food crop in their own right, but also the higher market prices that can be achieved for beans over maize and bean crop residues over maize stalks (as livestock fodder). The greater susceptibility of beans to pests and diseases is also cited as another reason why farmers plant more beans, so that they can be guaranteed some yield from this important crop.

Table: *Maize, Bean and Banana Varieties Grown in Arumeru*

Crop varieties	*Economic uses*	*Plant characteristics*
Zea mays (Maize)		
Kienyeji	Food, income, crop residuesfed to animals	Not very sweet, tolerant to storage pests, good milling quality, low yielding, drought susceptible.
Katumani	Food, income, crop residues fed to animals	Drought tolerant, early maturing, low yielding, good milling quality, tolerant to storage pests.
CG4141 (Lowlands)	Food, income, crop residues fed to animals	Good milling quality, drought tolerant.
UCA (Highlands)	Food, income, crop residues fed to animals	Good milling quality, drought tolerant.
Kilima	Food, income, crop residues fed to animals	High yielding, susceptible to storage pests, good milling quality, high water demand, high quality flour.
Phaseolus spp (Beans):		
Soya kijivu	Food, income, crop residues fed to animals.	"No gases after eating", early maturing, good taste, climbing type, sweet, high price, grey.
Kachina	Food, income, crop residues fed to animals.	High market price, early maturing, spoils quickly after cooking.
Lovirondo	Food and crop residues fed to animals	Climbing type, "causes bloating and gases after eating", laborious to harvest, low market price.
Bwanashamba	Food and crop residues fed to animals	Most popular in Kiserian, high yielding, good taste, susceptible to diseases and aphids.
Masai red ndogo (namira)	Food and crop residues fed to animals.	High yielding, good tasting, "no gases after eating", needs wide spacing for high production.
Karanga	Food and crop residues fed to animals.	High yielding, good tasting when cooked (flavours food).
Masai-red kubwa (namriri)	Food and crop residues fed to animals.	High market price, bush type, early maturing, good tasting and flavours food, susceptible to diseases.
Lyamungu 90	Food and income	Good tasting and flavours food, early maturing, drought tolerant, high yielding, high market price.
Kiburu	Food and crop residues fed to animals.	Drought tolerant, grows well on soils with poor fertility.
Engichumba	Food and income	Very high yielding, violet bean
Engichumba-ng'iro	Loshoro (traditional food)	High yielding, sweet, grey bean
Engichumba-narok	Food and income	Similar to Engichumba-ng'iro, black bean
Moshi	Food and income	Very high yielding, sweetest, yellow bean
Kibumulu	Food and income	Fast cooking, high price, dark red bean

Musa spp (Bananas)		
Kisimiti	Income, brewing, animal feed (stem)	Early maturing, drought tolerant, good milling quality
Ng'ombe	Loshoro, brewing, income, roofing, fodder to animals.	Hard when cooked.
Mshale	Matendela (traditional food), income	Good for roasting, long and thick banana fingers.
Uganda fupi	Banana soup (mtori), fruit, income, peels fed to animals	Early maturing, small with mainly fingers, susceptible to pests and diseases.
Uganda ndefu	Banana soup, fruit, peels fed to animals	Large with few fingers, susceptible to pests and diseases.
Kisukari	Fruit, income, animal feed (stems)	Very sweet, drought and disease tolerant, low nutrient demand
Mzuzu	Roasting for tea	Tolerant to drought and disease
Malindi	Food (matendela), animal feed	Drought tolerant
Mnanambu	Soup, roasting	Shade
Mkonosi	Roasting	Disease tolerant
Mkono wa tembo	Roasting	Disease tolerant
Ndishi	Loshoro, income	Susceptible to diseases
Olmuririko	Loshoro, brewing	Modest tolerance to diseases

Diversity at Landscape Level

It is typical in Ngiresi/Olgilai for farms to be made up of a number of different plots scattered throughout the village. The evolution of these farm types is explained first by the periodic clearance of more land as additional areas were brought into production and second by the practice of dividing farms between sons. In such divisions parents are inclined to split every plot into smaller plots to ensure that all sons receive land of similar quality.

Farmers in Ngiresi and Olgilai use different parts of the landscape in different ways, matching crop suitability to available land. In matching crops to plots of land, farmers not only consider biological suitability, but also the value of the crop, the labour required to manage the crop and the risk that the crop might fail.

Therefore, high value crops and fruit trees are planted close to the home to reduce the risk of theft, whereas hillside plots are planted with sweet potatoes to reduce soil loss from erosion.

Field borders are often planted with bananas and trees to demarcate boundaries. Farmers recognise the suitability of different plots to different cropping strategies.

For example, Mr. Yangan from Ngiresi/Olgilai, has nine different field types making up his farm. Some of these are detailed below.

Table: *Some Field Types on Mr. Yangan's Farm*

Location	***Field Type***	***Rationale***
Near house flat/gentle slopes	Tethering pasture	Convenient for milking morning and evening.
	House garden	Convenient to house
	Coffee/banana	Allows for easy transport of manure to the field. Also convenient for regular chores such as spraying coffee.
Middle distance gentle slopes	Maize/beans	Convenient enough for transporting inorganic fertilizers. In situ green manuring replaces animal manure.
Remoter steep slopes	Sweet potatoes Fodder grasses	Crops don't need a lot of inputs, but provide good cover for hillsides prone to erosion. Sweet potatoes in particular attract a lot of mice so it is preferred that these are not grown close to the house.
Boundaries	Bananas	Tree crops in particular are used to mark out boundaries.

Some of the older farmers in Kiserian have a deep understanding of the different soil types found in the village. Soils are classified locally by colour, fertility, depth and moisture holding capacity. Based on this categorisation crops are selected to match different soils and planting dates are determined.

In order to spread risk some plots and crops are dry planted, others are planted immediately after rain, while still others are planted a few days after the first rains. In these ways farmers are able to minimise the risks of poor crop yields due to low levels of soil fertility and moisture stress.

Diversity in Resource Management

Management practices reflect the value placed on the land and on the current crop by a farmer. Land tenure is often critical in determining how farmers manage their land. Secure tenure has been shown to be an incentive to careful stewardship of the land.

Open access property or fields held on short term rents are often less well tended. In the case of Kiserian the contrast in management practice is evident in the condition of communal and clan woodlots compared to those owned individually. The former are degraded while the latter are actively managed by the owner (for example, through the application of crop residues).

Table: *Soil management strategies under the major cropping systems in Olgilai/Ngiresi villages.*

Management Objective	*Cropping system*	*Soil Management Strategies*
Soil fertility improvement	Coffee/banana/maize/beans in rotation with round potatoes	FYM application, incorporation of crop residues, house refuse and weeds and application of ashes, planting of Sesbania sesban, grevillea and composting (few).
	Maize/beans	Incorporation of FYM, grevillea biomass, crop residues, green manuring, trash lines.
	Maize/beans rotation with sweet potatoes	Application of FYM, incorporation of crop residues and Sesbania biomass.
Soil moisture management	Coffee/banana/maize/beans in rotation with round potatoes	Incorporation of crop residues, mulching and trash lines, shading by coffee/banana/fruit trees/ canopy, planting of seteria.
	Maize/beans	Self-mulching, crop residues, trash lines, incorporation of crop residues (few).
	Maize/beans rotation with sweet potatoes	In-situ crop residues mulching, trash lines, sweet potato cover.
Soil erosion control	Coffee/banana/maize/beans in rotation with round potatoes	Rain interception by trees canopy, Mulching, trash lines, construction of flower hedges, trees canopy interception, planting of Sesbania sesban.
	Maize/beans	Trash lines, crop canopy
	Maize/beans rotation with sweet potatoes	Incorporation of crop residues, trash lines, application of ashes, planting of fodder grass strips.

Management practices also vary with landscape. Fertility management, in particular, differs depending on the spatial arrangement of a farmer's fields. In general, farmyard manure will be applied to fields close to the house/cow stall. Fewer farmers are likely to transport manure long distances for application on hillside plots. Instead they may use other biological means of enhancing soil fertility, or they may apply chemical fertilizers, which are easier to transport.

The following table details some of the soil management practices that are applied by farmers to fields in Olgilai/Ngiresi

In general, there is less emphasis in Kiserian on soil fertility management, with less use of both organic and inorganic fertilizers. Instead, in order to increase crop yields, farmers rely on bringing more land into cultivation, mixed cropping and variations in planting dates.

Agrodiversity – Expert Farmers

The PLEC approach is to identify 'expert' farmers and to facilitate the dissemination by them of their knowledge and experience. In this context expert farmers are those who have maximised the production potential on their farms in a sustainable and conserving way. In both Ngiresi/Olgilai and Kiserian this involves diversification and intensification of crop production. Gidiel Laizer farms six separate plots in Ngiresi/Olgilai. He plants different crops on different plots depending on their distance from his home, their topography and the other resources available to him (especially labour).

In 1998 Gidiel changed his planting strategy on a 0.25 ha plot surrounding his house. He now continuously crops this field. This plot is planted with the traditional perennial intercrop of coffee and bananas. However, Gidiel no longer plants the usual intercrop of maize and beans.

Instead, in response to the availability of local markets (Arusha town is just a few kilometres away), he plants cauliflower and round potatoes for sale in addition to maize for home consumption. This serial cropping system, yielding three crops per annum, is called a 'matatu' system and involves continuous cultivation of the plot. In the following table Gidiel's planting and harvesting calendar is summarised.

Table: *Matatu cropping system*

Month	***Activity***		
	Planting	***Harvesting***	***Other***
March	***Plant cauliflower***		***Manure applied***
April			
May	Plant round potatoes	Harvest cauliflower	
Junc			
July		Harvest round potatoes Harvest bananas	Manure applied
August	Plant maize	Harvest bananas	
Septembe r			
October			
Novembe r		Harvest coffee	Manure applied
Decembe r		Harvest coffee	
January			
February		Harvest maize	

In addition, further crops are planted along the plot boundary providing further sources of food which include taro (*Colocasia esculenta*), fodder (*Cariandra caryopsis*) and grevillea (*Grevillea robusta)* for which he sells timber to get cash. Gidiel is also planning to introduce yams (*Dioscorea spp*) into the border of the field as an additional source of food. High levels of diversity are supported on Gidiel's farm. In addition to the 'matatu' cropping system, which supports diverse crops in a single field, varietal diversity is also evident. Gidiel cultivates five different varieties of bananas and he is attempting to crossbreed two varieties of maize to develop his own strain combining the desirable traits from those that he is currently growing.

Gidiel selectively applies farmyard manure to the plots close to the house where he practises the 'matatu' system. He identifies those areas of the field where the yield was poor for the previous crop. Those parts of the field receive the first and greatest application of manure in order to improve their productivity under the next crop.

As a result of health concerns and for financial reasons, Gidiel does not use commercially produced chemical pesticides.

Instead he uses a mixture of fermented cattle urine (collected from the stall) and water, which he prepares himself.

Spraying this mixture on the plants also introduces additional nutrients to the crop. Another use of fermented urine (mixed with botanicals) is to control ticks in livestock.

The PLEC approach seeks to encourage the communication of information by farmers to farmers.

This is facilitated by farmer demonstrations, where farmers are invited to visit the farm of an 'expert' farmer who explains his practices.

In addition to the innovative matatu cropping system, Gidiel cultivates a large number of plants with traditional medicinal uses. Following demonstration meetings held at his farm, local farmers took away cuttings and seedlings from plants used in traditional medicine to propagate them on their own farms.

In Kiserian, the strategies followed by Lais Kitia, another expert farmer, differ from those of Gidiel Laizer, largely due to the different environments in which they farm.

Lais Kitia has established, over many years, a very successful agroforestry cropping system. In addition to cereals and legumes, Lais maintains a wide variety of useful trees, many of them local species, well adapted to the arid conditions. His agroforestry and cropping systems mean that he harvests a crop, be it cereal, beans or fruit, every few months.

The fruits are mostly harvested during the dry season and provide an important source of food for his family.

Because Kiserian is more remote from markets, Lais's production is focused on foodstuffs for home use rather than for sale in the market, although produce from his agroforestry system can be converted into cash if necessary.

Lais's farm has ten diverse field types. The crops grown in a single plot of less than 0.3ha are summarised in the following table. In addition to two annual crops (maize and millet), this field supports a wide variety of fruit trees and is used for tethering cattle and sheep.

Table: *Farming on One Field at Mr. Lais Kitia's Farm in Semi-arid Kiserian*

Type of plant	***Varieties***	***Economic uses***	***Plant characteristics***
Maize (*Zea mays*)	Malawi	Food, income, animal feed, firewood	High yielding, large grains, easy to mix with trees, good milling qualities.
Finger millet (*Eleusine indica*)	Enyangai (local)	Porridge, income, brewing, animal feed	High price, tolerant to drought and storage pests.
Mangoes (*Mangifera indica*)	Embedodo	Fruit, income	Large fruit, sweet, good aroma, good market price, keeps, green when ripe, drought tolerant.
	Embe mviringo	Fruit, income	Very sweet, fibrous, heavy, green when ripe, drought tolerant.
	Saforoni	Fruit, income	As sweet as honey, drought tolerant.
	Boribo	Fruit, income	Very sweet, high water content, orange and red when ripe, drought tolerant.
	Achari	Fruit, income, combines with chilli pepper, tomatoes, onion and coconut juice to make appetiser.	Very sweet, drought tolerant.
Pawpaws (*Carica papaya*)	Kienyeji 1	Fruit, income	Watery, very sweet, soft.
	Kienyeji 2	Fruit, income	Moderate water content, moderately sweet, hard.
Lemons (*Citrus limoni*)	Kienyeji	Income, porridge appetiser	Very sweet, drought tolerant.
Zambarau (*Syzygium guineense*)	Local	Fruit, income, shade.	Drought tolerant
Oranges (*Citrus cinensis*)	Local	Fruit, shade	Drought tolerant
Ukwaju (*Tamarindus indica*)	Local	Wild fruit, income	Drought tolerant, soil fertility improvement
Bananas (*Musa sapientum*)	Kisukari	Fruit, income, fed to animals.	Very small fingers, very sweet, drought tolerant.
	Kisimiti	Fruit, income, brewing, fed to animals.	Drought and disease tolerant
Mnafu (*Solanum nigrum*) (volunteer crop)	Local	Wild vegetable	Drought tolerant.
Wild amaranthus (*Amaranthus thunbergii*) (volunteer crop)	Local	Wild vegetable	Drought tolerant
Livestock	Varieties	Economic uses	Characteristics
Cattle (tethered on wild trees)	Local breed	Draught power, milk, manure, income, security, dowry, prestige.	Pests and diseases tolerant.
Sheep (tethered on wild trees)	Local	Meat, dowry, fatty foods to breast feeders.	Pests and diseases tolerant.

Conclusion and Policy Implications

Dryland farming systems in Africa display a remarkable resilience. If the past predictions of their demise were to have been true, we should be seeing far more environmental degradation and loss of biodiversity than we do today. Part of the reason for the resilience lies in the ability of farmers to adapt to changing conditions. The Boserupian hypothesis of adaptive change through the local application of technology and sustainable intensification is alive and well in many, but obviously not all, parts of Africa. This paper has sought to show that farmers in Arumeru, Tanzania, are no exception. They have developed some sometimes-intricate sets of techniques to manage their natural resources, they have used the biological diversity they have to hand, especially in regard to food crops, and they have organised their use of the whole landscape to secure their livelihoods and biodiversity.

Of particular relevance to developing lessons from these empirical findings are the ways in which farmers both understand and manage complex associations of plants, which then in turn provide food security to local households. Scientists and policy makers must be very careful not to write off the value in current practices of small-scale farmers as being of no relevance to today.

Debates about the unsustainability of dryland farming in Africa are rife with accusations about the degrading practices of small farmers. However, if land users have themselves developed sustainable and productive practices, then this is as worthy of publicity as the examples of bad practices. Bad news may sell newspapers; but good news should surely be equally prominent in giving policy makers a balanced picture.

The Arumeru farmers have developed an extraordinary diversity in crops and cropping systems. The PLEC project will be looking into what resources farmers need to access this rich harvest of information. However, it is already clear that poor farmers can, and do, practice sustainable use of biodiversity, and many use genotypes, varieties and landraces of crops that are now rare.

Food security for the many marginal populations in Africa, a crucial area of policy debate today, may well rest on these poor farmers' efforts. The Kiserian semi-arid site, for example, has bean varieties that withstand drought as well as provide both food to humans and crop residues to animals.

Multi-purpose varieties and plants that meet several needs are important for dryland farmers. If they also avoid – or minimise – the inherent riskiness of the climate, such plants are worthy of conservation.

They may not yield as highly as improved varieties, but for food security at household level this is not the key criterion.

The PLEC experience in Tanzania has also highlighted the beneficial, mutual interaction between professionals, field workers and farmers. Working closely with farmers on demonstration sites, our scientists have come to appreciate that they each have different kinds of knowledge, which are of equal importance. The "white-coat" syndrome that all that comes from scientific experimentation is superior – and by extension, all that comes from small-scale farmers and peasants is inferior – has bedevilled science. Our colleagues in PLEC have learnt to harmonise their experience with farmers and to attempt to interleave their knowledge from science with farmers' own experiences and advice. It has been a good mutual learning exercise for all involved. Policies on extension, agricultural education and interactions with clients (i.e. farmers) have much to gain.

Finally, PLEC in Tanzania has demonstrated the heterogeneity in rural society. There are innovative and expert farmers who are carrying on agricultural activities that have substantial scope for replication. These self-same farmers have training abilities also. Farmer-to-farmer extension has been a particular success in Arumeru, where field days and farmer-organised demonstrations have influenced many land users.

We have witnessed substantial change through such informal means of dissemination. We have seen women farmers teaching each other, swapping planting materials, and enthusiastically engaging in community works. By formally valuing the human resources at local level and appreciating the agrodiversity farmers have developed themselves, PLEC is empowering people and giving them a dignity that more traditional projects fail to emulate. Time will tell if the process will continue. Much will depend on political stability, social order and the continuing willingness of scientists and land users to collaborate. However, the signs are positive. Agrodiversity can help to support the global agenda of conservation of biodiversity, while at the same time providing for development of local people and the meeting of their needs.

Monsoon

A Monsoon of the Indian subcontinent is among several geographically distributed observations of global monsoon taking place in the Indian subcontinent. In the subcontinent, it is one of oldest weather observations, an economically important weather pattern and the most anticipated weather event and unique weather phenomenon.

Yet it is only partially understood and notoriously difficult to predict. Several theories have been proposed explaining the origin, the process, the strength, the variability, the distribution and the general vagaries of the monsoon of the Indian subcontinent, but understanding of the phenomenon and its predictability are still evolving. The unique geographical features of the subcontinent, along with associated atmospheric, oceanic and geophysical components, are extremely influential in ensuring the anticipated behaviour for a monsoon in South Asia and the Indian subcontinent. Due to its effect on agriculture, flora and fauna and the general weather of India, Bangladesh, Pakistan, Sri Lanka, etc., among other economic, social and environmental effects, a monsoon is one of the most anticipated, followed and studied weather phenomena of the Indian subcontinent.

It has significant impact on the overall well-being of subcontinent residents and has even been dubbed the "real finance minister of India".

Definition

Monsoon, derived from the Arabic word "Mawsim" meaning "season", although generally defined as a system of winds characterized by a seasonal reversal of its direction, lacks a consistent detailed definition. Some examples are given below:

- The American Meteorological Society defines it as *a name for seasonal winds*, first applied to the winds blowing over the Arabian Sea from northeast for six months and southwest for six months. Later it has been extended to similar winds in other parts of the world.
- Intergovernmental Panel on Climate Change (IPCC) describes Monsoon as a tropical and subtropical seasonal reversal in both the surface winds and associated precipitation, caused by differential heating between a continental-scale land mass and the adjacent ocean.
- Indian Meteorological Department defines it as the seasonal reversals of the wind direction along the shores of the Indian Ocean, especially in the Arabian Sea, that blow from the southwest during one half of the year and from the northeast during the other half.
- Colin Stokes Ramage in *Monsoon Meteorology*, International Geophysics Series, Vol. 15, defines Monsoon as a seasonal reversing wind accompanied by corresponding changes in precipitation.

Monsoon of the subcontinent is primarily noted for its rain bearing ability and for the associated unpredictability of the weather. Consequently some definitions incorporate rain in its definition.

Background

Observed initially by sailors in the Arabian sea travelling between Africa, India and South-East Asia, Monsoon is a major weather phenomenon in India (and the subcontinent) for the influence it casts on the lives of its inhabitants for centuries. Monsoon in India can be categorized into *two branches* based on their spatial spread over the sub-continent:

- Arabian Sea Branch
- Bay of Bengal Branch

Figure: *SW Monsoon clouds over Tamil Nadu*

Alternatively, it can be categorized into *two segments* based on the direction of rain bearing winds:

- South-West Monsoon (SW Monsoon)
- North-East Monsoon (NE Monsoon)

Based on the time of the year that these winds bring rain to India, they can also be categorised in *two rain periods* called:

- the *Summer monsoon* (May to September)
- the *Winter monsoon,* (October to November)

The complexity of Monsoon as a weather phenomenon of India is not yet completely understood, making it difficult to accurately predict its behaviour in terms of quantity, temporal and spatial distribution of the accompanying precipitation. These are the most monitored components of Monsoon determining the water availability in India for any given year.

Mechanism of Monsoon

Monsoon is a tropical phenomenon. Indian subcontinent, lies northwards of the equator up to the Himalayas and Hindukush, primarily in the tropical zone of the Northern Hemisphere. Weather pattern involves winds blowing from the south-west direction (known as South-West Monsoon) from the Indian Ocean onto the Indian landmass during the months of June through September. These are generally rain-bearing winds, blowing from sea to land, and bring rains to most parts of the subcontinent.

They split into two branches, the *Arabian Sea Branch* and the *Bay of Bengal Branch* near the southernmost end of the Indian Peninsula. They are eagerly awaited in most parts of India for their agricultural and economic importance.

Subsequently later in the year, around October, these winds *reverse* direction and start blowing from a north-easterly direction. Given their land to sea flow, from subcontinent onto the Indian Ocean, the system with less moisture brings rain to only limited parts of India like Kerala, Andhra Pradesh and Tamil Nadu.

This is known as the North-East Monsoon. However, this rain is responsible for the rice bowls of South India. This *mechanism* completes the *annual Monsoon cycle* of the Indian subcontinent.

Effect of Geographical Relief Features

Although the SW and NE Monsoon winds are seasonally reversible, they do not cause precipitation on their own.

Two factors for rains are essential for rain formation:

1. Moisture-laden winds
2. Droplet formation

Additionally, one of the causes of rain need to happen, which in this case is primarily Orographic due to presence of highlands right across the paths of the winds. Orographic barriers in the path of a wind force it to rise. Consequently, precipitation occurs on the windward side of highlands due to adiabatic cooling and condensation of the rising motion of the moist air.

For all the above scenarios to fulfill simultaneously, the unique geographic relief features of Indian subcontinent come into play. The notable features of Indian sub-continent, required in explanation of the Monsoon mechanism, are as follows:

1. First is the presence of abundant water bodies around the subcontinent - Arabian Sea, Bay of Bengal and Indian Ocean. These help in accumulation of moisture in the winds during the hot season.
2. Second is the presence of abundant highlands like the Western ghats and the Himalayas right across the path of the SW Monsoon winds. These are the main cause of the substantial orographic precipitation all over the Indian subcontinent.
3. The Western Ghats are the first highlands of India that the SW Monsoon winds encounter. The Western Ghats rise very abruptly from the Western Coastal Plains of the subcontinent making effective orographic barriers for the Monsoon winds.
4. The Himalayas play more than the role of just the orographic barriers for Monsoon. They help in its confinement onto the subcontinent. Without it, the SW Monsoon winds would blow right over the Indian subcontinent into China, Afghanistan and Russia without causing any rain.
5. For NE Monsoon, the highlands of Eastern Ghats play the role of orographic barrier.

Features of Monsoon Rains

There are some unique features about the rains that Monsoon brings to the Indian subcontinent.

"Bursting" *of Monsoon Rains*

Bursting of Monsoon implies the onset of the sudden change in weather conditions in India (typically from hot and dry weather to wet and humid weather during the SW Monsoon) due to abrupt rise in the *mean daily rainfall.*

Similarly the burst of NE monsoon marks an abrupt increase in the *mean daily rainfall* over the affected regions.

Monsoon Rain Variability ("Vagaries")

One of the most commonly used phrases to describe the erratic nature of the Monsoon of the Indian subcontinent is "vagaries of monsoon", used in newspapers, magazines, books, web-portals to insurance plans and India's budget discussions.

In some years, it rains too much causing floods in several parts of India, in others it rains too little or not at all causing droughts. In some years when the rain quantity is sufficient, its timing may be arbitrary. In some years, in spite of *average annual rainfall*, its daily distribution or the areal distribution might be substantially skewed. Such is the variability in the nature of Monsoon rains and weather.

Ideal and Normal Monsoon Rains

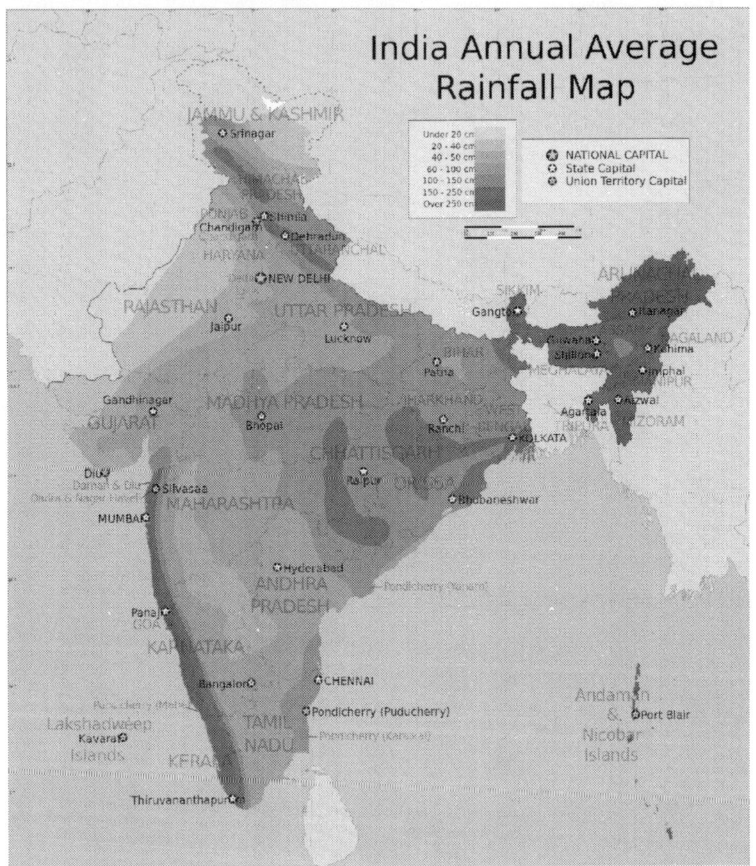

Figure: *Annual Average Rainfall Map of India*

Every year the normal onset of SW Monsoon is expected to *"burst"* onto the western coast of India (near Thiruvananthapuram) around 1 June covering entire India by around 15 July. Its withdrawal from India typically starts from 1 September onwards and completes by around 1 October.

Similarly the NE Monsoon is expected to *"burst"* around 20 October and last for a period of about 50 days before withdrawing.

However, a rainy Monsoon is not necessarily a *normal* Monsoon. A *normal Monsoon* is expected to perform close to its statistical averages calculated over a significantly long periods. Therefore, a *normal Monsoon* is generally accepted to be the Monsoon that has near *average quantity* of precipitation over all the geographical locations (*mean spatial distribution*) under its influence and over the entire expected time period of its influence (*mean temporal distribution*). Additionally, the *arrival date* and the *departure date* of both the SW and NE Monsoon should be close to the *mean* dates. The exact criteria for *Normal Monsoon* is defined by the Indian Meteorological Department with calculations for the *mean* and *standard deviation* for each of the aforesaid precipitation variables.

A Monsoon with excess rain can cause floods in India, Pakistan and Bangladesh. and one with too little rain can lead to widespread drought, food shortage, famine and economic losses. Therefore, a *normal Monsoon* with *mean performance* is the most desirable Monsoon.

Theories for Mechanism of Monsoon

Theories for mechanism of Monsoon primarily try to explain the reasons for the seasonal reversal of winds and the timings of their reversal.

Traditional Theory

Due to difference in the specific heat capacity of land and water, continents heat up faster than the seas. Consequently the air above the coastal lands heats up faster than air above seas. This creates areas of low air pressure above coastal lands compared to the air pressure over the seas causing winds to flow from the seas onto the neighbouring lands. This is known as sea breeze

Process of Monsoon Creation

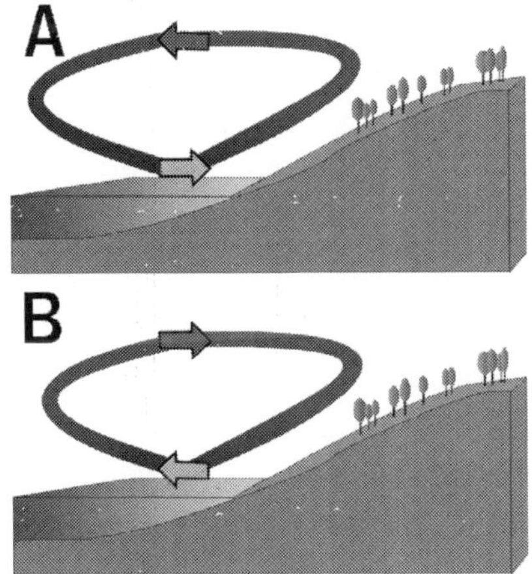

Figure: *A: Sea breeze, B: Land breeze*

Also known as the *thermal theory* or the *Differential Heating of Sea and Land Theory*, it portrays the Monsoon as a large-scale sea breeze. It states that during the hot sub-tropical summers, the massive landmass of Indian Peninsula heats up at a different rate than the surrounding seas resulting in a pressure gradient from South to North. This causes flow of moisture laden winds from sea to land. On reaching the land these winds rise up due to the geographical relief, cooling adiabatically and leading to orographic rains. This is the *southwest monsoon*. Reverse happens during winter when the landmass is colder than the sea establishing a pressure gradient from land to sea. This causes the winds to blow over Indian landmass towards Indian Ocean in a north-easterly direction causing the *northeast monsoon*. Since the SW monsoon is from sea to land, it has more moisture (therefore causing more rain) than the NE monsoon. Only a part of the NE monsoon passing over Bay of Bengal picks up moisture causing rain in Andhra Pradesh and Tamil Nadu during the winter months.

However many meteorologists argue that the Monsoon is not a local phenomenon as explained by the traditional theory but a general weather phenomenon along the entire tropical zone of earth. This criticism, does not deny the role of differential heating of sea and land in generating monsoon winds but merely restricts it to one of the several factors rather than the only one.

Dynamic Theory

The prevailing winds of the atmospheric circulation arise due to the difference in pressure at various latitudes of Earth and act as means for distribution of thermal energy on the planet. This pressure difference is due to the differences in Solar insolation received at different latitudes of Earth and the resulting uneven heating of the planet. Alternating belts of high-pressure and low-pressure develop along the equator, the two tropics, the Arctic and Antarctic circles and the two polar regions giving rise to the Trade winds, Westerlies and the Polar easterlies. However, the geophysical factors like revolution of earth, its rotation and axial tilt of the Earth result in gradual shifting of these belts northwards and southwards following the Sun's seasonal shifts.

Process of Monsoon Creation

The *dynamic theory of Monsoon* explains monsoon on the basis of the annual shifts in the position of global belts of pressure and winds. According to it, Monsoon is the result of the shift of the Inter Tropical Convergence Zone (ITCZ) under the influence of the vertical sun. Though the mean position of the ITCZ is taken as the equator it keeps shifting northwards and southwards with the migration of the vertical sun towards the tropics (Tropic of Cancer and Tropic of Capricorn) during the summer of the respective hemispheres (Northern and Southern Hemisphere). As such, the theory states that during the northern Summer (months of May and June), the ITCZ moves northwards, along with the vertical sun, towards the Tropic of Cancer. The ITCZ being the zone of lowest pressure in the tropical region, is the target destination for the Trade winds of both the hemispheres. Consequentially, with ITCZ at the Tropic of cancer, the South East Trade winds of the Southern Hemisphere have to cross the equator to reach the ITCZ. However, due to Coriolis effect, (Coriolis effect causes winds in northern hemisphere to turn to its right whereas winds of southern hemisphere to turn to its left) these South East trade winds are deflected eastwards in the Northern Hemisphere transforming into South West trades. These pick up the moisture while travelling from sea to land and cause orographic rain once they hit the highlands of the Indian Peninsula. This results in the South-West Monsoon.

The *dynamic theory* provides the explanation of the system of Monsoon as a circum-global weather phenomenon rather than just a local one. And when coupled with the *Traditional Theory* (based on heating of Sea and Land) it enhances the explanation of the differential intensity of precipitation impact of Monsoon along the coastal regions with orographic barriers.

Jet Stream Theory

This theory tries to explain the establishment of both the NE and SW Monsoons as well their unique features like *bursting* and variability. The jet streams are a system of upper-air westerlies. It gives rise to slowly moving upper-air waves, with 250 knots winds in some air streams. First observed by World War II pilots, they develop just below the tropopause over areas of steep pressure gradient on the surface. The main types are the *polar jets*, the *subtropical westerly jets* and the less common *tropical easterly jets*. They follow the principle of geostrophic winds.

Process of Monsoon Creation

Over India, a *subtropical westerly jet* develops in the winter season which is replaced by the *tropical easterly jet* in the summer season. The high temperature over the Tibetan Plateau, as well as over Central Asia in general, during the summer is believed to be the critical factor leading to the formation of the *tropical easterly jet* over India in summer. The mechanism affecting monsoon is that the *westerly jet* causes high pressure over northern parts of the subcontinent during the winter.

This results in the north to south flow of the winds in the form of the NE Monsoon. With the northwards shift of the vertical sun, this jet shifts northwards too. The intense heat over the Tibetan Plateau, coupled with associated terrain features of high altitude of the plateau, etc. generate the *tropical easterly jet* over central India. This jet creates a low pressure zone over the northern Indian plains influencing the wind flow towards these plains, assisting the establishment of the SW Monsoon.

Theory for "Bursting" of Monsoon

The unique feature of *bursting* of the Monsoon is primarily explained by the *Jet Stream theory* and the *Dynamic Theory.*

Dynamic Theory

According to this theory, during the summer months of Northern Hemisphere, the ITCZ shifts northwards pulling the SW Monsoon winds onto the land from the sea. However the huge landmass of the Himalayas continue to restrict the low pressure zone onto the Himalayas itself. It is only when the Tibetan Plateau heats up a lot more than the Himalayas does the ITCZ abruptly and swiftly shift northwards leading to *burst of Monsoon showers* over the Indian subcontinent. The reverse shift takes place for the NE Monsoon winds leading to a second minor *burst* during the Northern Hemisphere winter Months of NE Monsoon rainfall over Eastern Indian peninsula.

Jet Stream Theory

According to the theory the onset of SW Monsoon over Indian subcontinent is driven by the shift of the *subtropical westerly jet* northwards from over the plains of India towards the Tibetan Plateau. This shift is due to the intense heating of the Plateau during the summer months. This shift of the *westerly jet* to the north of the Himalayas is not a slow and gradual process, as expected for most changes in weather pattern.

The primary cause of these is believed to be the height of the Himalayas. As the Tibetan Plateau heats up the low pressure created over it pulls the *westerly jet* northwards. Due to lofty Himalayas, the *westerly jet* is inhibited from moving northwards. However, with continuous dropping pressure, sufficient force is created for the movement of the *westerly jet* across the Himalayas after a significant period. As such, the shift of the jet is sudden and abrupt causing the *bursting of SW Monsoon rains* onto the Indian plains. The reverse shift happens for the NE Monsoon.

Theories for Monsoon Variability

The Jet Stream Effect: The above mentioned Jet Stream theory also explains the variability in *timing* and *strength* of the Monsoon.

Timing: A timely northward shift of the *subtropical westerly jet* in the beginning of the summer season is critical to the onset of the SW Monsoon over India. If the northward shift of this jet is delayed, so is the SW Monsoon. An early shift heralds in an early Monsoon.

Strength: Additionally, the strength of the SW Monsoon is determined by the strength of the *easterly tropical jet* over central India. A strong *easterly tropical jet* results in a strong SW Monsoon over central India and vice-versa.

El Niño-Southern Oscillation (ENSO) Effect

El Niño is a 'warm' ocean current originating along the coast of Peru that replaces the usual 'cold' Peru or Humboldt Current. This warm surface water reaching towards the coast of Peru with El Niño is pushed westwards by the trade winds thereby raising the temperature of the southern Pacific Ocean. A reverse condition is known as La Niña.

Southern Oscillation, a phenomenon first observed by Sir Gilbert Thomas Walker Director-General of Observatories in India, refers to the seesaw relationship of atmospheric pressures between Tahiti and Darwin, Australia. He noticed that when it was high pressure in Tahiti, it was low pressure in Darwin and vice versa. A *Southern Oscillation Index* (SOI),based on the pressure difference between Tahiti and Darwin, has been formulated by the Bureau of Meteorology (Australia) to measure the strength of the Oscillation. Walker noticed that the quantity of rainfall in the Indian subcontinent was often negligible in the years of high pressure at Darwin (and low pressure at Tahiti). Conversely, low pressure at Darwin bode well for the precipitation quantity in India. Thus he established the relationship of Southern Oscillation with quantities of Monsoon rains in India.

Ultimately, it was realized that the Southern Oscillation is just the corresponding atmospheric component of the El Niño/La Niña effect (which happens in the Ocean). Therefore in the context of the Monsoon, the two cumulatively came to be known as the *ENSO*. The ENSO is known to have a pronounced effect on the strength of SW Monsoon over India with the Monsoon being weak (causing droughts in India) during the El Niño years whereas La Niña years had particularly good Monsoon strength over India.

Indian Ocean Dipole Effect

Although ENSO was statistically effective in explaining several past droughts in India, in the recent decades the ENSO-Monsoon relationship seemed to weaken in the Indian subcontinent. For e.g. the 1997, strong ENSO failed to cause drought in India. However, it was later discovered that just like ENSO was an event in he Pacific Ocean, a similar seesaw ocean-atmosphere system in the Indian Ocean was also at play. It was discovered in 1999 and named the Indian Ocean Dipole (IOD).

An index to calculate it was also formulated. IOD develops in the equatorial region of Indian Ocean from April to May peaking in October. With a positive IOD winds over the Indian Ocean blow from east to west. This results in the Arabian Sea (western Indian Ocean near African Coast) being much warmer and eastern Indian Ocean around Indonesia becoming colder and dry.

In the negative dipole year, reverse happens making Indonesia much warmer and rainier. It was demonstrated that a positive IOD index often negated the effect of ENSO, resulting in increased Monsoon rains in several ENSO years like the 1983, 1994 and 1997. Further, it was shown that the two poles of the IOD - the eastern pole (around Indonesia) and the western pole (off the African coast) were independently and cumulatively affecting the quantity of rains for the Monsoon in the Indian subcontinent.

Equatorial Indian Ocean Oscillation

Similar to ENSO, the atmospheric component of the IOD was later discovered and named as *Equatorial Indian Ocean Oscillation (EQUINOO).* When EQUINOO effects were factored into the statistics certain failed forecasts, like the acute drought of year 2002, could be further accounted for. The relationship between extremes of the Indian summer monsoon rainfall, along with ENSO and EQUINOO have been studied and models for enhanced predictability of the quantity of monsoon rains have been statistically derived.

Monsoon Rain Prediction Models

Since the Great Famine of 1876–78 in India, various attempts have been made to predict the rainfall during the Monsoons in India. At least five models for prediction of Monsoon rains exist in India.

Seasonal Prediction of Indian Monsoon (SPIM)

Centre for Development of Advanced Computing (CDAC) at Bengaluru is facilitating the Seasonal Prediction of Indian Monsoon

(SPIM) experiment on the PARAM Padma supercomputing system. This project did simulated runs on historical data from 1985 to 2004 to try and establish the relationship between of five *atmospheric general circulation models* with the Monsoon rainfall distribution.

Indian Meteorological Department Model

IMD has tried to forecast the Monsoon for India since 1884, some unsuccessfully but till 2011 is the only official agency entrusted with making public forecasts about the quantity, distribution and timings of the Monsoon in India.

IMD's position as the sole authority on the matter was further reiterated in 2005 by the Department of Science and Technology (DST), New Delhi. In 2003, IMD underwent a substantial change in its forecast methodology, model as well as its administration. A sixteen parameter monsoon forecasting model used by the Indian met office since 1988 was revamped in 2003 with a new one. However, following the 2009 drought in India (worst since 1972), IMD decided in 2010 that it needed to develop an *indigenous model* to further enhance its prediction capabilities.

Significance

Indian Monsoon is the primary delivery mechanism for fresh water in the Indian subcontinent. As such it impacts the environment (and associated flora, fauna and ecosystems), agriculture, society, hydro-power production and geography of the subcontinent (like availability of fresh water in water bodies, underground water table) with all these factors cumulatively contributing towards the health of the economy of affected countries.

The Indian Monsoon turns large parts of India from a kind of semi-desert into green grasslands.

Geographical (Wettest Spots on Earth)

Mawsynram and Cherrapunji, both in the Indian state of Meghalaya alternate to be the *wettest places on Earth* given the quantity of their rainfall. There are other cities with similar claims but with 10,000mm of rain for each of these locations, Monsoon of Indian subcontinent is a significant contributor towards the water supply of its area of influence.

Agricultural

India, historically an agrarian economy primarily, has recently seen the services sector overtaking the farm sector in terms of GDP contribution. However, even today agriculture sector contributes 17-20% of GDP and is the largest employer in the country with about 60% of people dependent on it for employment and livelihood. The land use pattern of India indicates that 49% of land is under agriculture in India, it is 55% if associated wetlands agriculture, dryland farming areas, etc. are included.

Since over half of these farmlands are rain-fed, Monsoon is critical to the food sufficiency and quality of life for the country. Despite progress in alternative forms of irrigation, agricultural dependency on monsoon is far from insignificant, even today.

Therefore, the agricultural calendar of India is governed by Monsoon. Any fluctuations in the time distribution, spatial distribution or quantity of the monsoon rains may lead to conditions of floods or droughts causing the agricultural sector to adversely suffer. This has a cascading effect on the secondary economic sectors, the overall economy, food inflation and therefore the overall quality and cost of living for the general population in India.

Economic

The economic significance of monsoon can be aptly summed up by Pranab Mukherjee's statement that monsoon is the *real finance minister of India.* A good monsoon resulting in improved agricultural brings down prices of essential food commodities and reduces their imports overall reducing the food inflation. Further improved rains result in increased hydroelectric production. All these factors initiate positive ripple effects throughout the economy of India.

Social

D. Subbarao (Governor of Reserve Bank of India), during a quarterly review of the monetary policy, once highlighted that lives of Indians depends on the performance of Monsoon. His own personal career prospects, emotional well being and the performance of his monetary policy were all *a hostage to monsoon* like it was for most Indians. Additionally, farmers, rendered jobless due to failed Monsoon rains tend to migrate city-wards. This crowds the city slums and further aggravates the job, infrastructure and sustainability of city life. Such is the magnitude of effect that monsoon casts on the lives of Indians.

Travel

In past, people usually refrained from travelling during monsoons for practical as well as religious reasons. But with advent of globalization, travel during monsoons is gaining popularity. Places like Kerala, Western

Ghats get a very large number of tourists, both local and foreigner during monsoon season. Kerala is one of the top destinations for tourists interested in Ayurvedic treatments and massage therapy. One major drawback of travelling during monsoons is the fact that most of wildlife sanctuaries are closed during this rainy season. Also, some mountainous areas, specially in Himalayan regions get cut off due to damaged roads caused by landslides and floods during heavy rains.

Environmental

The Monsoon is the primary bearer of fresh water to bodies of water in the area. The peninsular/Deccan rivers of India are mostly rain-fed and non-perennial in nature depending primarily on the Monsoon for water supply. Similarly most of the coastal rivers of Western India are rain-fed and Monsoon dependent. As such, obviously the flora, fauna and the entire ecosystem of these areas are heavily dependent on the Monsoon.

Dryland Agriculture and Conservation Technologies for Improving Soil Moisture for Crop Production in Semiarid Kenya

Dryland agriculture is practised where there is water stress and seasonal agricultural drought due to low water supply (low rainfall, high runoff water losses and high evaporation). In dryland agriculture, emphasis is on conservation of soil water for crop production. Dryland agriculture is defined as land husbandry under conditions of moderate to severe soil moisture stress during a substantial portion of the year, which requires special cultural techniques, adapted crops and sustainable production systems. Typical characteristics of dryland areas include: low total rainfall with at least one pronounced dry season; highly variable and unreliable rainfall during the normal rainy season with temporal (year to year, month to month) variations in rainfall amount and distribution; increase in persistence of agricultural drought with decreasing rainfall; potential evapo-transpiration exceeding rainfall for at least seven months of the year; and very high intensity rainstorms of short duration which often lead to high surface runoff and soil erosion. Thus limited soil water supply and a serious erosion potential are the two most common characteristics in all dryland areas.

The three broad areas of soil, water and nutrient management in drylands include: management of soil water; management of soil erosion; and the management of fertilization. Less soil water could lead to some water stress. Likewise less fertilization could lead to some nutrient stress. Severe soil erosion would result in the loss of nutrient rich topsoil.

In dryland areas, the most dominant soil types - Lixisols, Luvisols, and Acrisols (FAO/UNESCO Classification) are of limited agricultural productivity due to soil related problems. Thus the four major problems limiting dryland crop production are severe soil erosion, low soil fertility, low soil moisture and high soil crusting and compaction. These four soil management problems are stratified according to slope, agro-climatic zones, land use and soil type. The occurrence of soil crusting and compaction is attributed to seasonal rainfall characteristics (amount, intensity and duration), physical soil properties and bad tillage practices. The problem of soil compaction is caused by bad tillage practices (leading to the development of plough pans) and natural soil conditions [subsurface hardpans due to clay eluviation (leaching) from the A to B horizons]. Soil crusting and compaction decreases rainwater infiltration and increases surface runoff. Increased and concentrated surface runoff water flow causes severe soil erosion. Of the four soil related problems, Low soil moisture is the most critical as it directly affects dryland agriculture.

The effectiveness of soil, water and nutrient management interventions in dryland areas could be objectively assessed in terms of: enhanced infiltration of rainwater; increased soil water storage capacity; improvement of effective use of stored water; minimization of evaporation from the soil; and minimization of soil loss and salinization.

In order to realize the above objectives and enhance soil and crop productivity, there is need to increase rainfall and crop water use efficiency (RWUE and CWUE) and fertilizer use efficiency (FUE). These are verifiable indicators of effective soil and water management in dryland areas. Rainfall water use efficiency could be improved through: runoff waterharvesting and conservation (RWHC) for crop and fodder production; crop rotation where deep rooted plants would use water in deeper soil horizons; surface residue mulching and cover cropping to decrease surface evaporation; enhancement of infiltration and reduction in surface runoff; and ultimately an increase in crop water use efficiency. An increase in crop water use efficiency could be realized through: enhanced transpiration; application of fertilizer and organic manure; and enhancement of drought tolerance by crops. Likewise a high fertilizer use efficiency when there is some good crop response due to availability of nitrogen, phosphorus and potassium.

In semi arid Kenya, rainfall is bimodal, and characterized as low, erratic and poorly distributed. The short and long rainy seasons receive about 55% and 45% of the total annual rainfall respectively. The short rains (October to December) are more reliable, evenly distributed and adequate for crop production. The long rains (March to May) are associated with most crop failures due to the poor distribution, unreliability and inadequacy for crop production. The dominant soil types e.g. Lixisols, Luvisols, and Acrisols (FAO/UNESCO Classification) differ widely in depth, texture, chemical fertility, organic matter content and physical properties. In this area, soil crusting and compaction is a common soil management problem and is attributed to structural instability of the dominant soils (Luvisols, Lixisols, and Acrisols) and bad tillage practices. The problem of soil compaction in Iiuni is caused by bad tillage practices (leading to the development of plough pans) and natural soil conditions [subsurface hardpans due to clay elluviation (leaching) from the A to B horizons].

So what are the on-farm conservation tillage technologies that have been tested under dryland conditions. The tillage methods that have been tried include zero tillage, conventional tillage with two farmyard manure (5 and 10 Mg/ha). Farmyard manure (FYM) when incorporated into the soil enhanced infiltration and reduced soil crusting and compaction, and surface runoff during the initial stages of a rainy season. In the mid-stages of the rainy season, the effects of soil aggregation from FYM diminished and this resulted in increased soil loss. Towards the end of the rainy season when soil crusts had formed, some significant decrease in soil loss with FYM treatments was observed. Conventional tillage without farmyard manure led to high surface runoff and soil loss for Luvisols (with inherent low aggregate stability). Zero tillage has performed poorly under these soil conditions because of the significantly high soil crusting and compaction, surface runoff and subsequent low rainwater infiltration.

Recently (1999-2001) three tillage techniques that included conventional tillage, subsoiling and ripping/ridging and pitting/ridging were tried out through on farm demonstration trials in a semi arid area of Kenya (Biamah, 2001). Conventional tillage was done using the conventional ox-drawn mouldboard plough to a maximum depth of 10 cm. Below this depth, is a plough pan that often developed due to compaction from continuous conventional tillage operations. Subsoiling and ripping/ridging was done on fallow land using an animal drawn subsoiler and a ripper/ridger.

The subsoiling was done in strips along planting rows to a maximum depth of 25 cm. The ripper/ridger was used to increase the width of cut to about 5-6 cm up to the tillage depth of 25 cm and this created plain ridges for increased depression storage of surface runoff. Pitting and ridging was done on fallow land.

The pits were manually dug using hand hoes. Each pit was dug to a diameter of 10 cm and depth of 20 cm. The labour involved was quite intensive and costly. In each pit, 2 to 3 handfuls of farmyard manure (FYM) were applied and mixed with the topsoil before planting. Some depression was maintained in each pit to impound runoff water during the rainy period. From the statistics obtained during the short rains (2000/2001) and the long rains (2001), it was evident that pest (white grub and termite) damage accounted for about 18% to 28% of the total crop losses in Iiuni, Machakos.

Table: *Plant population, crop damages and percent losses due to termite activity during the short rainy season, 2000-2001 at Iiuni, Machakos, Kenya (Biamah, 2001).*

CONTIL technique	*Plant population (No)*	*Crop damage (No)*	*Crop loses (%)*
Conventional tillage	990	275	28
Animal subsoiling and ripping	965	265	27

Table: *Plant population, crop damages and percent losses due to white grub and termite activity during the long rainy season, 2001 at Iiuni, Machakos, Kenya (Biamah, 2001)*

CONTIL technique	*Plant population (No)*	*Crop damage (No)*	*Crop loses (%)*
Conventional tillage	714	141	20
Animal subsoiling and ripping/ridging	772	128	17

- The crop yields presented in Tables, show a trend of good grain and biomass yields during the short rainy seasons of 1999 and 2000. Under these semi arid rainfall conditions, these yields are above normal.
- The maize grain yields obtained by conservation tillage farmers during the short rainy season of 2000 were the highest in several decades of their farming career.
- The crop yields for the long rainy seasons (2000 and 2001) were good when compared with what a conventional farmer would get. During the two long rainy seasons, no conventional farmer harvested any maize grain.

Table: *Average grain and biomass yields during the short rainy season, 1999-2000 at Iiuni, Machakos, Kenya (Biamah, 2001).*

CONTIL technique	*Grain yield (kg/ha)*	*Biomass yield (kg/ha)*
Conventional tillage	1700	2560
Animal subsoiling and ripping	2580	3080

Table: *Average grain and biomass yields during the short rainy season, 2000-2001 at Iiuni, Machakos, Kenya. (Biamah, 2001).*

CONTIL technique	*Grain yield (kg/ha)*	*Biomass yield (kg/ha)*
Conventional tillage	2893	3268
Animal subsoiling and ripping	3121	3290

Table: *Average grain and biomass yields during the long rainy season, 2000 at Iiuni, Machakos, Kenya. (Biamah, 2001).*

CONTIL technique	*Grain yield (kg/ha)*	*Biomass yield (kg/ha)*
Pitting and ridging	2875	2300

Table: *Average grain and biomass yields at Iiuni, Machakos, Kenya. Long rainy season, 2001 (Biamah, 2001).*

CONTIL technique	*Grain yield (kg/ha)*	*Biomass yield (kg/ha)*
Conventional tillage	342	1268
Animal subsoiling and ripping/ridging	385	1529

Farmers Perceptions

Farmer experiences with conventional tillage practices in the trials area showed that improvements in soil and crop productivity occur only when the soil erosion problem is minimized first through effective conservation structures. Thus conservation tillage practices are complimentary to erosion control and provide remedies to soil fertility, soil moisture and soil compaction problems. Thus conservation tillage is viewed as a viable land management option for improving soil moisture conservation for dryland crop production in semi arid Kenya.

Conclusions

- In the conservation tillage trials area, conventional tillage operations using the ox-drawn mouldboard plough resulted in the development of plough pans below a maximum ploughing depth of 10 cm. Besides there was the natural subsurface hard pan that is attributed to clay eluviation in the B_t horizon.
- The breaking of subsurface hardpans calls for conservation tillage techniques that improve the depth of tillage to 25~30 cm and thus enhancing the soil moisture storage in the soil. The depth of tillage in the trials area varied with the tillage implements used and soil conditions especially the antecedent soil moisture conditions.

- Deep or shallow tillage could be recommended on the basis of seasonal soil conditions as influenced by rainfall characteristics. For instance, deep and shallow tillage are recommended for the short and long rainy seasons respectively. Deep tillage is an effective tillage operation for enhancing soil moisture storage during the long dry spell before planting. Likewise shallow tillage is an appropriate tillage operation for minimizing soil moisture losses through evaporation.

Dryland Cropping Systems

Quick Facts...

- Soils under no-till production systems store more water than soils on conventional stubble mulch systems and allow conversion to more intense crop rotations.
- Conversion to no-till systems using a three- to four-year rotation such as wheat-corn-fallow, wheat-millet-fallow, or wheat-corn-millet-fallow, results in more profitability.
- Conversion to no-till, three- to four-year rotations results in greater risk and requires greater management input.

The Wheat-Fallow (WF) system, practiced for many years in the semiarid Western Great Plains, was a definite improvement over continuous cropping. It stabilized yields and provided farmers with a reliable income from year to year. However, wheat fallow is an inefficient user of annual precipitation in a region where water is the major limiting factor. Data collected in the Great Plains show that typically less than 25 percent of the precipitation received during the 14-month fallow period is stored in the soil. When little or no residue is left on the soil surface, water storage efficiency is less than 20 percent.

Colorado and Nebraska research indicates that precipitation storage efficiencies of 40 to 60 percent is achieved when tillage is minimized or eliminated. The key to these improvements is maintaining crop residue on the soil surface and minimizing or eliminating soil disturbance. Unfortunately, little added wheat yield results from the additional water stored under no-till compared to mulch tillage systems. In fact, it costs more to save the additional water in a no-till system than the value of the added grain yield. Weed control with herbicides is more costly than with mechanical tillage.

Water savings in no-till systems can be converted to profit only by switching to more intense cropping systems where fallow time is decreased and summer Crops, like corn, grain sorghum, proso millet and annual forages are added to the rotation.

Maximizing water storage, which permits more intensive rotations, requires residue maintenance on the soil surface to trap snow, absorb raindrop impact, slow runoff and minimize evaporation, as well as complete weed control at critical times in the production system.

Producers can use the following recommendations to improve precipitation use efficiency in eastern Colorado dryland farming systems, and improve soil, water and air quality while increasing net income. These steps are built around rotations where winter wheat is the starting point in the system.

Weed Control: Weed control in growing wheat is the goal of this step. If controlled in the wheat, the stubble at harvest is essentially weed free. Weeds include winter annuals such as mustard, kochia and Russian thistle.

If weeds are present at harvest, apply the appropriate herbicide for that species. Apply residual herbicides in late July to mid-August to control volunteer wheat, downy brome and jointed goatgrass. Various combinations of herbicides with small amounts of atrazine are effective for this purpose.

Keep the atrazine rates low and within labelled amounts for your soils and climate. Failure results in long-term damage to wheat yields.

Do not till in this phase of the rotation. Dollars spent for herbicides in this phase give maximum returns compared to any other point in the system. It should be possible to store 3 to 5 inches of water between wheat harvest to fall freeze-up if the stubble is kept weed free. By not tilling, the maximum amount of standing stubble can trap snow in the winter months and increase water storage by spring.

Spring Crop Choices and Cultural Practices: Possible summer Crops to follow wheat in the rotation are corn, sorghum, proso millet, sudex or sunflowers. Plant corn in areas north of Cheyenne Wells and grain sorghum in southern areas. Proso millet is a good option in the northern area, but not in the southern areas where yields and markets are poor. Annual forages, like sudex and hay millet, are excellent choices for livestock producers.

Whatever the summer crop, plant it into the wheat stubble or other residue base with minimal disturbance. A minimum till or slot planter is recommended. Apply an appropriate herbicide according to the label and use for weed control during the summer period. Avoid cultivation of row Crops unless the herbicide fails. Choose crop varieties that adapt to your elevation, latitude and longitude.

Chapter 7

Biophysical Aspects of Carbon Sequestration in Drylands

The process of CS, or flux of C into soils forms part of the global carbon cycle. Movement of C between the soil and the atmosphere is bidirectional. Consequently, carbon storage in soils reflects the balance between the opposing processes of accumulation and loss. This reservoir of soil C is truly dynamic. Not only is C continually entering and leaving the soil, the soil C itself is partitioned between several pools, the residence times of which span several orders of magnitude. Nor is soil C an inert reservoir, the organic matter with which it is associated is vital for maintaining soil fertility and it plays a part in such varied phenomena as nutrient cycling and gaseous emissions. A detailed description and analysis of soil C and organic matter can be found elsewhere (Schnitzer, 1991; FAO, 2001c). Taking into account specific biophysical characteristics of dry areas.

Halophytes

A special feature of many dryland soils is salinity, either through natural occurrence or increasingly as a result of irrigation. Saline soils affect large parts of the drylands (Glenn *et al.*, 1993). Such lands are often abandoned, but halophytic plants are especially adapted to these conditions and offer potential for sequestering C in this inhospitable environment. It has been estimated that 130 million ha are suitable for growing halophytes, which can be used for forage, feed and oilseed. Glenn *et al.* estimate that 0.6 - 1.2 gigatonnes of C per year could be assimilated annually by halophytes. Evidence from decomposition experiments suggests that 30 - 50 percent of this C might enter long-term storage in soil. Although irrigation would be required to achieve these figures, a complete carbon budget still suggests a carbon actual rate of 22 - 30 percent.

Grasslands

Grasslands are the natural biome in many drylands, partly because rainfall is insufficient to support trees, and partly because of prevailing livestock management. However, grassland productivity and CS have been controversial. The productivity of tropical grasslands is now known to be much higher than was previously thought, and consequently they sequester much more C (Scurlock and Hall, 1998). Estimates for C stored under grassland are about 70 tonnes/ha, which is comparable with values for forest soils. Although many of the grassland areas in drylands are poorly managed and degraded, they offer potential for CS as a consequence.

The average annual input of organic matter into grassland is about double the 1 - 2 tonnes/ha that is contributed to cropped soils (Jenkinson and Rayner, 1977). This fact is borne out by the results of studies from various locations. The data have shown that grassland, even where subject to controlled grazing, generally has higher soil C levels than cropland. Chan and Bowman (1995) found that 50 years of cropping soils in semiarid New South Wales, Australia, had on average reduced soil C by 32 percent relative to pasture. The reduction was linearly related to the number of years of cropping. Similarly, soils of tall-grass pasture under controlled grazing had greater soil C than adjacent cropland subject to conservation tillage (Franzluebbers *et al.*, 2000).

The key factor responsible for enhanced carbon storage in grassland sites is the high carbon input derived from plant roots. It is this high root production that provides the potential to increase SOM in pastures and vegetated fallows compared with cropped systems. Root debris tends to be less decomposable than shoot material because of its higher lignin content (Woomer *et al.*, 1994). Consequently, the key to maintaining and increasing CS in grassland systems is to maximize grass productivity and root inputs (Trumbmore *et al.*, 1995). Grasses have also been shown to sequester more C than leguminous cover crops (Lal, Hassan and Dumanski, 1999). Grasses also have the potential to sequester C on previously degraded land. Garten and Wullschleger (2000) used a modelling approach that estimated a 12-percent increase in soil C could be obtained under switchgrass (*Panicum virgatum* L.) on degraded land in ten years.

Grazing is a feature of many grasslands, natural or managed. This might be expected to decrease the availability of residues that can be used to sequester C, especially as the quantity of C returned in manure is less than that consumed. However, provided there is careful grazing management, many investigations have found a positive effect of grazing on the stock of soil C. This was found to be the case for a composite pasture (alfalfa and perennial grasses) in the semi-arid pampas (Diaz-Zorita, Duarte and Grove, 2002). Even under harsher conditions in the Syrian Arab Republic, grazing was found to have no detrimental impact on soil C (Jenkinson *et al.*, 1999). Schuman, Janzen and Herrick, (2002) have calculated that with proper grazing management, rangelands in the United States of America can increase soil carbon storage by 0.1 - 0.3 tonnes/ha/ year. For new grasslands, this can rise to 0.6 tonnes/ha/year.

The positive effect of grazing appears to result from the effect that it has on species composition and litter accumulation. Willms *et al.* (2002) found that when prairie was protected from grazing there was little effect on production but there was an increase in the quantity litter. Reeder and Schuman (2002) also found that there was an accumulation of litter in an ungrazed semi-arid system and that soil carbon levels were higher in the grazed lands. The litter acted as a store of immobilized C. The ungrazed grassland also experienced an increase in species that lacked the fibrous rooting system that is conducive to SOM formation and accumulation.

Therefore, grasslands can play a vital role in sequestering C. However, careful grazing management is essential. The historical record shows how susceptible semiarid grasslands are to overgrazing, soil degradation and carbon loss.

Burning

Fires form part of the natural cycle in many biomes and are especially prevalent in grassland ecosystems. However, humans have also used fire to clear areas for agriculture and to clear crop residues. The action of fire would seem to be counter to CS as it returns C fixed by vegetation directly to the atmosphere, thereby preventing its incorporation into the soil. The effect of fire is difficult to generalize because it depends upon the intensity and speed of the fire. These factors are influenced by: the state of the vegetation, i.e. its maturity and woody component; the accumulation of litter; and climate factors such as moisture level. They C in all the aboveground material that is fully combusted will be lost from the system. However, in grassland ecosystems, C lost to fires can be replaced quickly by increased photosynthesis and vegetative growth (Knapp, 1985; Svejcar and Browning, 1988). Even in savannah systems that contain woody species, it has been shown that C lost through combustion can be replaced during the following growing season (Ansley *et al.*, 2002). Regarding the soil, the intensity and speed of the fire will govern the depth to which it is affected. In one study where burning was used to clear forests, 4 tonnes C/ha was lost in the top 3 cm of soil, but this was replaced within one year under a pasture system (Chone *et al.*, 1991).

Not all plant material is fully combusted by fire, and a variable amount of charcoal will be produced. Charcoal is extremely resistant to decomposition; it is not cycled like most organic matter, and has a mean residence time of 10 000 years (Swift, 2001). Consequently, in severely degraded soils that have been affected by several fires, charcoal and charred material can form a substantial proportion of the remaining organic C. However, it is not known whether charcoal and other charred material have any protective effect on the native organic matter. Thus, although fires release CO_2 back into the atmosphere, the simultaneous production of charcoal can be regarded as a sequestration process that results in a substantial amount of C being accumulated within the soil for a long period.

Afforestation

Forestry is recognized as major sink for C. However, as well as accumulating C aboveground, it can also make significant contributions to soil C even in drylands. A number of suitable species are available that enable viable afforestation in dryland environments (Srivastava *et al.*, 1993; Silver, Ostertag and Lugo, 2000; Kumar *et al.*, 2001; Niles, *et al.*, 2002). In particular, N-fixing trees generally lead to increased accumulation of soil C. For example, *Prosopis* and *Acacia* are adapted to subtropical semi-arid lands and are reported to increase the level of soil C by 2 tonnes/ha (Geesing, Felker and Bingham, 2000).

Some tree varieties are particularly suitable for growing on degraded lands, their deep rooting systems tapping resources unavailable to shallow-rooted crops. *Prosopis juliflora* has been grown on salt-affected soils in northwest India and increased the SOC pool from 10 tonnes/ha to 45 tonnes/ha in an eight-year period (Garg, 1998). Even where large-scale forestry is not appropriate, there is often scope for introducing trees around farmers' fields. Such is the case in semi-arid India where *Prosopis cineraria* has improved soil fertility as well as sequestering extra C (Nagarajan and Sundaramoorthy, 2000). However, natural systems are complex and trees do not guarantee improved CS. Jackson *et al.* (2002) found that soil C decreased when woody vegetation invaded grasslands. Although there was an increase in aboveground and belowground biomass, these gains were offset by losses in soil C.

Residues

Plant residues provide a renewable resource for incorporation into the SOM. Production of plant residues in an ecosystem at steady-state will be balanced by the return of dead plant material to the soil. In a native prairie, about 40 percent of plant production is accumulated in the SOM (Batjes and Sombroek, 1997). However, in agricultural systems, because plants are harvested, only about 20 percent of production will on average be accumulated into the soil organic fraction. Furthermore, in some farming systems, all aboveground production may be harvested, leaving only the root biomass. Of the plant residue returned to the soil, about 15 percent will be converted to passive SOC (Lal, 1997). Schlesinger (1990) is more pessimistic, suggesting that only 1 percent of plant production will contribute to CS in soil. The actual quantities of residue returned to the soil will depend on the crop, the growing conditions and the agricultural practices. For example, for a soybean - wheat system in subtropical central India, the annual contribution of C from aboveground biomass was about 22 percent for soybean and 32 percent for wheat (Kundu *et al.*, 2001). This resulted in 18 percent of the annual gross carbon input being incorporated into the SOM. In semi-arid Canada, the conversion of residue C to SOC was reported to be 9 percent in frequently fallowed systems, increasing to 29 percent for continuously cropped systems (Campbell *et al.*, 2000).

Unless a root crop is being harvested, all belowground production is available for incorporation into the SOM. Roots are believed to be the major constituent of particulate organic matter although tillage reduces the net accumulation of C from roots substantially (Hussain, Olsson and Ebelhar, 1999). In cool climates, belowground carbon inputs from roots alone can generally maintain soil carbon levels. However, this is not the case in warmer or semi-arid regions where residues are decomposed much more readily, providing sufficient moisture is available (Rasmussen, Albrecht and Smiley, 1998). Consequently, when continuous cropping is practised in drylands, failure to return aboveground plant residue will lead invariably to a reduction in soil C. Many African soils demonstrate this phenomenon. Continuous cropping over a number of years without the recommended inputs has often halved their carbon content (Woomer *et al.*, 1997; Ringius, 2002).

Both the quality and quantity of plant residues are important factors for determining the amount of C stored in soil. The quantity is highly dependent on the environmental conditions and agricultural practices. Differences between crops can be marked. A crop of maize will return nearly twice as much residue to the soil compared with soybean and, consequently, will result in a higher rate of SOM increase (Reicosky, 1997). The advantage that cereals have over legumes for achieving maximum CS rates has also been demonstrated by Curtin *et al.* (2000). They have shown that while black lentil fallow in semi-arid Canada added between 1.4 and 1.8 tonnes C/ha, a wheat crop would add 2 - 3 times this amount of C annually. Similarly, in Argentina, soybean, which produced 1.2 tonnes/ha of residue, resulted in a net loss of soil C, while maize, with 3.0 tonnes/ ha of residue, lessened the loss of soil C from the system significantly (Studdert and Echeverria, 2000).

Even within one crop group, large differences in organic matter production occur. Abdurahman *et al.* (1998) compared dry leaf production from pigeon pea and cowpea. While the former yielded 3 tonnes/ha, cowpea produced 0.14 tonnes/ha. These examples illustrate how the choice of crop can have a major influence on how much C an agricultural system can sequester.

The importance of roots relative to shoots for providing soil C is another factor illustrated in an experiment to compare the fate of shoot- and root-derived C (Puget and Drinkwater, 2001). In this study with leguminous green manure (hairy vetch), nearly half of the root-derived C was still present in the soil after one growing season whereas only 13 percent of shoot-derived C remained. This implies that shoot residues are broken down rapidly on account of their higher N content (Woomer *et al.*, 1994) and may serve as a nitrogen source for the following crop.

The chemical composition of plant residues affects their rate of decomposition. On average, crop residues contain about 40 - 50 percent of C but N is a much more variable component. A high concentration of lignin and other structural carbohydrates together with a high C:N ratio will decrease the rate of decomposition. For example, measurement of CO_2 evolution from tree leaves of African browse species and goat manure showed a significant correlation with initial nitrogen content and a negative correlation with lignin content (Mafongoya, Barak and Reed, 2000). Legume residues such as soybean are generally of high quality (low C:N ratio) and so decompose rapidly (Woomer *et al.*, 1994). Although the chemical composition of the plant residue affects its rate of decomposition, there is little effect on the resulting SOM (Gregorich *et al.*, 1998).

Where residues accumulate on the soil surface, their physical presence affects the soil. Mulches reduce water loss and soil temperature (Duiker and Lal, 2000), both important factors for drylands, especially where the soil temperature is above the optimum for plant growth. The ability of soil to assimilate organic matter is not clear-cut. A linear relationship between application and accumulation of SOM is often quoted. However, measurement of CO_2 flux in central Ohio, the United States of America, showed this flux to increase with added application of wheat straw (0, 8 and 16 tonnes/ha). However, the SOC was 19.6, 25.6 and 26.5 tonnes/ha after four years, suggesting that CS was reaching saturation (Jacinthe, Lal and Kimble, 2002).

Care must be taken when applying residues, as large losses of C can still occur under certain conditions. For example, in western Kenya, 70 - 90 percent of the added C was lost within 40 d when green manure from agroforestry trees was applied during the rainy season (Nyberg *et al.*, 2002). In Niger, the addition of millet residue and fertilizer for five years had no significant effect on the carbon level in the sandy soils (Geiger, Manu and Bationo, 1992). Termite activity may have contributed to the low levels of soil C here because entire surface mulches can be consumed within one year.

The potential amount of residues available for applying to soils can be large. Gaur (1992) estimated that in India about 235 million tonnes of straw/stover are produced annually from five major cereals (not only from drylands). Even if half of this were used for feeding livestock, there would be more than 1 million tonnes available for adding to the soil. However, availability of sufficient plant residues is often a problem, especially where they are required for livestock feed. This conflict of requirement occurs frequently in many dryland farming systems. In West Africa, crop residues are either removed or burnt. Consequently, the amount of soil C has declined steadily. Continuous cultivation and manure application can raise soil carbon levels by 40 percent but this frequently involves mining C from neighbouring areas to support the livestock (Ringius, 2002). Where plant and animal residues are in short supply, possibilities may exist for alternative organic inputs to the soil. For example, in India, waste products from a plant processing coir dust have been incorporated successfully into soil (Selvaraju *et al.*, 1999), and other experiments with industrial glue waste (Dahiya, Malik and Jhorar, 2001) have increased soil carbon levels successfully.

With regard to the complete carbon budget, where plant residues accumulate *in situ*, there is no extra carbon cost involved. Consequently, the C fixed by plants in photosynthesis is available as a net gain to the soil. The situation becomes more complex where machinery is required to separate the residue from the harvestable components. Where plant residues are transported between fields the energy cost would need to be included.

Perhaps the major issue regarding residue application and carbon accounting is the question of whether C is simply being transferred from one place to another rather than being truly sequestered. If organic material from industrial processes or other sources were to be used for incorporation into the soil, then any C used in transport would have to be accounted for. However, where the material is truly a waste product, there should be no need to consider the C used in its production. In this case, the C available for sequestration should be viewed no differently from the CO_2 emitted from a fossil fuel source that is subsequently fixed by plants in photosynthesis and returned to the soil via crop residues.

Applied Manures

The application of FYM has long been treated as a valuable source of organic matter to enhance soil fertility. One of the key characteristics of manure application is that it promotes the formation and stabilization of soil macroaggregates (Whalen and Chang, 2002) and particulate organic matter (Kapkiyai *et al.*, 1999). Manure is more resistant to microbial decomposition than plant residues are. Consequently, for the same carbon input, carbon storage is higher with manure application than with plant residues (Jenkinson, 1990; Feng and Li, 2001). Following five years of application, soil receiving manure had 1.18 tonnes/ha more C present than soil receiving plant residues. Even after 15 years, there was a still a difference of 0.37 tonnes C/ha as calculated by the RothC soil carbon model. In the field, Gregorich *et al.* (1998) found that manured soils had large quantities of soluble C with a slower turnover rate than in control or fertilized plots.

The composition and, therefore, decomposition of manure varies between the species from which it originates and also within species according to their diet (Somda and Powell, 1998). Many field trials have found that manure is the best means for incorporating organic matter into soils and promoting carbon storage. For example, Li *et al.* (1994) found that manure yielded the largest amount of C sequestered over a range of soils and climate conditions, although soil texture was important, and the greatest rate of sequestration occurred where there was a high clay content. However, many traditional agrosystems add manure in combination with fertilizers (Haynes and Naidu, 1998).

Depending on the system, the application of even relatively high amounts of FYM does not guarantee an increase in soil C. A long-term study in Kenya has shown that SOM declined even when manure was applied and maize residue returned (Kapkiyai *et al.*, 1999). It has been estimated that in order to maintain soil C in this system, 35 tonnes/ha of manure or 17 tonnes/ha manure with 16 tonnes/ha of stover would be required annually (Woomer *et al.*, 1997). Consequently, there is a carbon shortfall in this case, but it is difficult to see how the present system could remedy it. In addition, high application rates of manure can sometimes cause problems in the soil through the accumulation of K^+, Na^+ and NH_4^+ and the production of water-repellent substances by decomposer fungi (Haynes and Naidu, 1998). An additional problem in drylands that restricts the quantity of manure that can be applied is "burning" of the crop when insufficient moisture is available at the time of application. Consequently, farmers often wait until the rains have come before making an application, especially as precipitation is often erratic in arid regions. is often erratic in arid regions.

The production of sufficient manure for application to fields is a real problem for many smallholder farming systems. In Nigeria, direct manure input onto land from dry-season grazing is about 111 kg/ha of dry matter (Powell, 1986). This quantity will have little effect on the soil. More useful is the practice of night-parking cattle as manure production is usually greatest at dawn and dusk. When 50 cattle were penned in an area of 0.04 ha for five nights, they produced the equivalent of 6.875 tonnes/ha of manure (Harris, 2000). Normally, cattle will be penned in fields for 2 - 3 nights in northern Nigeria and this can supply manure at a rate of 5.5 tonnes/ha. Alternatively, in densely populated parts such as the Kano closed-settlement zone, cattle and crop production are fully integrated. Cattle are kept permanently in pens and fed using feed grown on neighbouring fields. Their manure is collected and spread onto the croplands. Although an efficient system, some C will be lost as a consequence of the respiratory and growth requirements of the cattle. An additional problem associated with cattle rearing is that ruminants produce significant quantities of CH_4, which is a potent GHG.

There has been some debate as to the usefulness of animal manure in CS. Schlesinger (1999, 2000) has calculated that providing 13.4 tonnes/ha would require 3 ha of cropland in order to produce sufficient cattle feed. This means that manure production requires mining of C from neighbouring lands. Although this is a generalization, it is argued that the 3:1 differential makes it unlikely that manure production per se could be used as a means of providing a net carbon sink in soils. However, in many dryland smallholder cropping systems, manure application rates are much lower, and fodder production is also less efficient.

Smith and Powlson (2000) have made the case that keeping cattle is part of many agricultural systems and, hence, that manure should be considered as a byproduct that can be added to arable land without necessarily including the carbon cost of its production. Part of the disagreement with Schlesinger on the usefulness of manure depends on where the system boundary for carbon accounting is drawn. However, when conducting such a carbon audit, it is essential to remember that the purpose of agriculture is to feed people; offsetting GHGs can only be evaluated as a secondary activity.

Inorganic Fertilizers and Irrigation

Fertilization and irrigation are primary means for increasing plant production and crop yield. Any increase in biomass also offers increased scope for CS. Consequently, irrigation and fertilization have been recommended as, and proved to be, successful methods of increasing CS (Lal, Hassan and Dumanski, 1999). Rasmussen and Rohde (1988) have shown a direct linear relationship between long-term nitrogen addition and accumulation of organic C in some semi-arid soils in Oregon, the United States of America. However, these technologies provide no additional organic matter themselves but do carry a carbon cost. Schlesinger (1999, 2000) has pointed out that pumping water requires energy and that the process of fertilizer manufacture, storage and transport is energy intensive. Consequently, Schlesinger (2000) has estimated that the gains in C stored using either fertilization or irrigation are offset by losses elsewhere in the system. Irrigation can also lead to the release of inorganic C from the soil.

Izaurralde, McGill and Rosenberg (2000) have argued that the calculations used by Schlesinger are based on very high rates of fertilizer application. For many dryland agriculture systems in the developing world, farmers do not have sufficient funds to apply large quantities of fertilizer even if they were available. With regard to the energy costs incurred in pumping water, solar-powered systems are being developed (Sinha *et al.*, 2002) and dryland environments with frequently clear skies would be able to make the best use of solar radiation.

These examples serve to illustrate how important it is to consider the whole system when CS is being considered to offset CO_2 emissions. The exact carbon cost of irrigation and fertilization would require calculation for each system, but the carbon deficits associated with both technologies make their incorporation into net CS systems difficult to achieve. Water conservation, the growing of legumes and careful nutrient cycling are more likely to yield a positive carbon balance.

Tillage

Pretty *et al.* (2002) consider tillage to be one of the major factors responsible for decreasing carbon stocks in agricultural soils. Research and experimentation with reduced tillage practices are most prevalent in the Americas. The mould-board plough and disc harrow are believed to be the causes of the loss of soil C through the destruction of soil aggregates and the acceleration of decomposition by the mixing of plant residues, oxygen and microbial biomass. Soil aggregates are vital for CS (Six, Elliott and Paustian, 2000), a process that is maximal at intermediate aggregate turnover (Plante and McGill, 2002). Of the organic matter fraction, the particulate organic matter is the most tillage sensitive (Hussain, Olsson and Ebelhar, 1999).

It is difficult to quantify the effects of tillage on soil C because the effect are very site dependent, e.g. coarse-textured soils are likely to be more affected by cultivation than are fine ones (Buschiazzo *et al.*, 2001). However, reducing tillage should be most effective in hot, dry environments (Batjes and Sombroek, 1997).

Reicosky (1997) conducted an experiment that used measurements of CO_2 efflux to investigate tillage-induced carbon loss from soil. The flux of CO_2 was monitored for 19 d following different forms of tillage practice. The mould-board plough buried most of the crop residue and produced the maximum CO_2 flux. The C released by the different treatments as a percentage of C in the crop residue was: 134 percent with mould-board plough; 70 percent with mould-board plough and disc harrow; 58 percent with disc harrow; 54 percent with chisel plough; and 27 percent with no-tillage. This demonstrates the correlation between CO_2 loss and tillage intensity, and demonstrates why farming systems that use mould-board ploughing inevitably lose soil C. Very large amounts of organic matter would be required to replace the loss incurred by such heavy tillage. Reicosky *et al.* (1995) estimate that 15 - 25 tonnes/ha manure plus crop residue would be needed annually in North America to offset these losses.

The flux of CO_2 from soil generated directly by the tillage process may not always reflect the overall release of CO_2 and hence carbon storage of the system. This is illustrated by a comparison of conventional disc tillage and no-tillage in central Texas, the United States of America, by Franzluebbers, Hons and Zuberer (1995). Here, seasonal evolution of CO_2 was up to 12 percent greater in the no-tillage system after 10 years. This was despite the fact that more C was sequestered by the no-tillage system. The authors suggest that a change in the dynamics of CS and mineralization have occurred under the no-tillage system. Similarly, Costantini, Cosentino and Segat (1996) found that more CO_2 was released from no-tillage or reduced-tillage compared with conventional tillage despite there being increased levels of soil C. They ascribe this difference to an increase in the microbial biomass.

Rates of carbon loss through tillage depend considerably on the site and cropping system. Ellert and Janzen (1999) measured the flux of CO_2 following the passage of a heavy cultivator on a semi-arid Chernozem soil in the Canadian prairies. They found that although tillage increased rates of CO_2 loss by two to four times, values returned to normal after 24 hours. They calculated that even with ten passes of the cultivator, only 5 percent of crop residue production would be released from this cropping system. In another situation, ploughing of a wheat - fallow cropping system near Sydney, Australia, reduced soil C by 32 percent after 12 years. Elimination of ploughing and adopting a no-tillage approach was unable to prevent a decrease in the carbon stock, although the loss was reduced to just 12 percent (Doran, Elliott and Paustian, 1998). The authors suggest that a fallow period would be required to halt the decline in soil C at this site.

There are many different types of tillage system. Conservation tillage covers a range of practices - no-tillage, ridge-tillage, mulch-tillage (Unger, 1990). Mulch-tillage maintains higher levels of residue cover. With mulches, only a small fraction of the residue is in contact with the soil surface and the microbes it contains. Decomposition is slow, especially as oxygen availability is limited. The physical presence of crop residues on the soil surface also alters the microclimate of the upper soil layer, which tends to be cooler and wetter compared with conventional tillage (Doran, Elliott and Paustian, 1998).

The accumulation of residues also reduces the loss of CO_2 from the soil surface. Alvarez *et al.*, (1995) reported an increase in labile forms of organic matter under no-tillage in the Argentine rolling pampa, indicating a decrease in the mineralization of the organic fraction. This study also noted that although organic C increased by 42 - 50 percent under no-tillage compared with ploughing and chisel tillage, there was also a marked stratification in the distribution of C under the no-tillage regime that was not evident in the ploughed system.

Stratification of organic C is common with reduced or no-tillage. Zibilske, Bradford and Smart (2002) in semi-arid Texas, the United States of America, demonstrated that the organic carbon concentration was 50 percent greater in the top 4 cm of soil of a no-tillage experiment compared with ploughing, but the difference dropped to just 15 percent in the 4 - 8-cm depth zone. This is typical of organic carbon gains observed with conservation tillage in hot climates. Bayer *et al.* (2000), working on a sandy clay loam Acrisol, also found that the increase in total organic C was restricted to the soil surface layers under no-tillage but that the actual quantity depended on the cropping system. An oat/

vetch - maize/cowpea no-tillage system produced the largest quantity of crop residues and sequestered the most C: 1.33 tonnes C/ha/year in 9 years.

Reicosky (1997) has compared the results from many no-tillage trials. The data emphasize the effect that crop rotation and quantity of crop residue has on organic matter accumulation. Overall, rates of organic matter accumulation can be expected to be lower in the hotter climates. Nevertheless, even in the very sandy soils of in north of the Syrian Arab Republic, it has been possible to make modest increases in SOM with no-tillage (Ryan, 1997). In western Nigeria, no-tillage combined with mulch application had a dramatic effect, increasing soil C from 15 to 32.3 tonnes/ha in 4 years (Ringius, 2002).

Although no-tillage systems are an excellent tool for combating the carbon losses associated with conventional cultivation, they do have their own special problems. In temperate lands, the reduction in soil temperature commonly associated with plant residue accumulation on the soil surface can retard germination. However, in drylands, where soil temperatures are frequently above the optimum for germination and plant establishment, such cooling is likely to be beneficial (Phillips *et al.*, 1980). No-tillage systems frequently suffer from an increased incidence of pests and diseases; the mouldboard plough and disc harrow are efficient weed controllers (Reicosky *et al.*, 1995). Consequently, no-tillage systems generally rely upon extra herbicides and pesticides. These inputs have an economic price and they also incur a carbon cost. However, in many dryland farming systems of the developing world, purchasing such products is not feasible, and quite often there is plentiful labour available for weeding. Application of N fertilizer can also be problematic when used on undisturbed, no-tillage soils. Where the soil is poorly drained, denitrification can occur and the reduced rate of evaporation increases the risk of nitrate leaching. In addition the native soil N has a lower rate of mineralization in undisturbed soil.

Not all soils are suited to a reduced-tillage approach. Some soils in the Argentine pampa may actually lose more C under no-tillage (0.7 - 1.5 tonnes C/ha/year) compared with conventional ploughing (Alvarez *et al.*, 1995) and periodic ploughing is required to avoid soil compaction (Taboada *et al.*, 1998). Where no-tillage is used on the pampas, the physical status of the soil is a critical factor for the success of the system (Diaz-Zorita, Duarte and Grove, 2002). Similarly, in the West African Sahel, the highest crop yields are obtained with deep ploughing, which is required to prevent crusting and alleviate compaction. In general, the success of reduced-tillage systems is often dependent on soil texture (Needelman *et al.*, 1999).

A particular advantage of the no-tillage system is that it favours multicropping; harvesting can be followed immediately by planting (Phillips *et al.*, 1980). Any cropping system that allows for continuous or near continuous plant growth should yield the maximum capacity to produce plant biomass and, consequently, has the potential to provide the greatest amount of organic mater for inclusion into the soil.

Considering the overall carbon budget, no-tillage systems have a lower energy requirement because tillage is very energy intensive. Phillips *et al.* (1980) have calculated that no-tillage systems in North America reduce the energy input into maize and soybean production by 7 and 18 percent, respectively. Improved water-use efficiency means that the energy, and hence carbon, cost of irrigation are reduced. However, the impact of energy savings is frequently offset by additional herbicide requirements (Phillips *et al.*, 1980). Kern and Johnson (1993) estimated that the manufacture and application of herbicides to no-tillage systems of the Great Plains is equivalent to 0.02 tonnes C/ha.

Reduced-tillage systems were adopted originally to help combat soil degradation. They were not intended as a means of sequestering C, which is a fortunate side-effect. Although the effectiveness of no-tillage at sequestering C will depend on the specific agricultural system to which it is introduced, there is no doubt that, as the intensity of tillage decreases, the balance between carbon loss and gain swings toward the latter.

Rotations

The importance of rotation in agricultural systems has long been known and the procedure now forms an intricate part of many conservation tillage practices. The inclusion of rotations has many benefits such as countering the buildup of cropspecific pests and, thereby, lessening the need for "carbon costly" pesticides and herbicides. Different crop species have a variety of rooting depths and this aids in distributing organic matter throughout the soil profile. In particular, deep-rooting plants are especially useful for increasing carbon storage at depth, where it should be most secure. The inclusion of N-fixing varieties in a rotation increases soil N without the need for energy-intensive production of N fertilizers.

The beneficial effects that rotations have for CS have been proved in many longterm field experiments. For example, Gregorich, Drury and Baldock (2001) made a comparison of continuous maize cultivation with a legume-based rotation. The rotation had a greater effect on soil C than did fertilizer.

The difference between monocultured maize and the rotation was 20 tonnes C/ha while the effect of fertilization was 6 tonnes C/ha after 35 years. In addition, the SOM present below the ploughed layer in the legume-based rotation appeared to be more biologically resistant. This demonstrates that soils under legume-based rotations tend to preserve residue C. A positive effect on SOC (an increase of 2-4 tonnes/ha) was also found with legumes and alternate cattle grazing in semi-arid Argentina (Miglierina *et al.*, 2000).

Rotations, especially legume-based ones, are generally regarded as extremely valuable for maintaining soil fertility and have a very good potential for sequestering C in dryland systems. Drinkwater, Wagoner and Sarrantonio (1998) estimate that their use in the maize/soybean-growing region of the United States of America would increase soil CS by 0.01 - 0.03 Pg C/year. The effectiveness of rotations for sequestering C is likely to be greatest where they are combined with conservation tillage practices.

Fallows

The role that fallows play in CS is varied. Where cropping is not taking place, it is important that vegetation cover is preserved. This is especially so in drylands where exposed soil is most likely to suffer from erosion and degradation. In addition to protecting the soil, cover crops can utilize solar energy that would otherwise be wasted. The CO_2 fixed is then available for sequestering into the soil as the plants senesce. The importance of vegetation cover can be illustrated with the results from an experiment conducted at a semi-arid site in Mediterranean Spain (Albaladejo *et al.*, 1998). Four and a half years after the vegetation cover was removed from one site, the SOC had decreased by 35 percent compared with the control plots.

The type of fallow is important. In Nigeria, forest clearance caused a decline in soil C from 25 to 13.5 tonnes/ha in seven years, but 12 - 13 years of bush fallow restored the carbon content (Juo *et al.*, 1995). Conversely, pigeon pea fallow was unable to sequester sufficient C on account of its low biomass production and rapid degradation.

However, fallows can have a negative effect on carbon storage in many situations. The frequency of summer fallows in semi-arid regions has been suggested as one of the major factors influencing the level of soil C in agricultural systems (Rasmussen, Albrecht and Smiley, 1998). Reducing the summer fallow in the semi-arid northwest United States of America is reported to have had a more positive effect on soil carbon retention than that achieved by decreasing tillage intensity. The loss of C in this region is believed to reflect the high rates of biological oxidation that occur here, which can only be offset by very large applications of manure (Rasmussen, Albrecht and Smiley, 1998). Consequently, yearly cropping and the associated organic additions is the recommended practice. Miglierina *et al.* (2000) also found that reducing summer fallow increased soil C, a consequence of the additional crop residue that was added. Using the CENTURY agro-ecosystem model, Smith *et al* (2001) predicted that reducing summer fallow in wheat cropping systems (wheat - fallow to wheat - wheat - fallow) in the semiarid Chernozems of western Canada would reduce carbon losses by 0.03 tonnes/ha.

Elimination of fallows can be highly beneficial for soil C simply because most fallows are associated with small inputs of plant residue. The significance of fallows for CS in a given system will depend on whether or not the cropping cycle adds significant quantities of organic matter to the soil. Where it does, then the presence of fallows is unlikely to enhance carbon storage within the system. Conversely, where the cultivation practice is poor and little or no organic matter is added, fallow periods will serve to counter this situation.

Soil Inorganic Carbon

Not all soil C is associated with organic material; there is also an inorganic carbon component in soils. This is of particular relevance to drylands because calcification and the formation of secondary carbonates is an important process in the soils of arid and semi-arid regions where, as a result, the largest accumulations of carbonate occur (Batjes and Sombroek, 1997). The dynamics of the inorganic carbon pool are poorly understood although it is normally quite stable. Sequestration of inorganic C occurs via the movement of HCO^-_3 into groundwater and closed systems. According to Schlesinger (1997) accumulation of calcium carbonate is quite low at 0.0012 - 0.006 tonnes/ha. However, Lal, Hassan and Dumanski (1999) believe that the sequestration of secondary carbonates can contribute 0.0069 - 0.2659 Pg C/year in arid and semi-arid lands.

Although soil inorganic C is relatively stable, it will release CO_2 if the carbonates become exposed through erosion (Lal, Hassan and Dumanski, 1999). In addition, irrigation can cause inorganic C to become unstable if acidification takes place through inputs of N and sulphur. The release of CO_2 through carbonate precipitation is seen as a major problem if irrigation waters are used in any system that is trying to store C. Furthermore, Schlesinger (2000) has pointed out that the groundwater of arid lands often contains up to 1 percent calcium and CO_2. This concentration is much higher than that which occurs in the atmosphere. Consequently, when these waters are applied to arid lands, CO_2 is released to the atmosphere and calcium carbonate precipitates. Schlesinger's calculations suggest that irrigation of some cropping systems would yield a net transfer of CO_2 from the soil to the atmosphere.

Trace Gases

An important aspect of agricultural systems in relation to the global carbon balance is the production of trace gases, particularly CH_4 and N_2O. When CS by soils is being considered as a mechanism for offsetting greenhouse gas emissions, it is necessary to consider all the interacting factors that can influence global warming. Both CH_4 and N_2O are radiatively active gases and, like CO_2, contribute to the greenhouse effect. Although they are present in the atmosphere at much lower concentrations than CO_2, they are much more potent. CH_4 and N_2O are, respectively, 21 times and 300 times more active GHGs than CO_2.

Ruminants, composting, biomass burning and waterlogging produce CH_4, while N_2O is released from soils when N fertilizer or manure is applied (Vanamstel and Swart, 1994). Manure usage is considered to be the major problem with regard to trace-gas emissions in agriculture. This is a potentially serious problem because the application of manure is a major tool for increasing soil C in drylands. Smith *et al* (2001) calculated that, for European soils, the effect of trace gases is sufficient to reduce the CO_2 mitigation potential of some no-tillage and manure-management practices by up to a half. Moreover, climate change is likely to amplify the problem as increased temperature is predicted to promote N_2O emissions (Li, Narayanan and Harriss, 1996).

Climate Change

Climate is a major factor involved in soil formation. Consequently, climate change will influence soils. Photosynthesis and decomposition will be affected directly and, hence, have an impact on soils. Whether soil carbon levels increase or decrease will depend on the balance between primary production and decomposition (Kirschbaum, 1995). Overall, productivity is predicted to increase as a consequence of rising CO_2 concentration and temperature, and this will lead to increased amounts of residue available for incorporation into the soil. However, higher temperatures can be expected to increase mineralization of SOM because this process is more sensitive to temperature increases than primary production. Kirschbaum (1995) predicts that SOC stocks will decline overall with global warming. However, Goldewijk *et al.* (1994)

suggest that the effects of temperature and water availability on soil respiration will be smaller than those attributable to the CO_2 fertilization effect. The direction of change is not certain but the balance of change will most probably operate at the regional level.

It has been argued that agricultural systems are to some extent buffered from environmental effects, while decomposition is not protected (Cole *et al.*, 1993). Hence, increased rates of mineralization might be more significant than any enhancement of production. However, the quality of plant organic matter is expected to decrease under elevated CO_2 owing to an increase in the C:N ratio. This would slow the rate of degradation (Batjes and Sombroek, 1997). Globally, the drylands are expected to become moister (Glenn *et al.*, 1993), which should lead to an increase in productivity and decomposition. However, shifts in climate zones are dependent upon a complex array of variables. Predictions based on the CENTURY agro-ecosystem model suggest that, overall, grasslands will lose soil C except in tropical savannahs, which should show a small increase (Parton *et al.*, 1995). Experiments at elevated CO_2 have also shown that changes in soil C in agro-ecosystems are particularly dependent upon the crop species grown (Rice *et al.*, 1994).

The full extent of the global rise in temperature associated with climate change may not be felt in many of the drylands because warming is predicted to be greatest at higher latitudes. With regard to the vegetation, some of the best-adapted plants for dryland regions use the C4 photosynthetic pattern. Because these species already have a CO_2-concentrating mechanism, they show little or no increase in productivity at elevated CO_2. However, they are still likely to receive some benefit from the increased water-use efficiency that accompanies a rising CO_2 concentration.

Case Studies on Drylands

Most of the research and case studies on soil CS have been conducted in temperate zones. Much less work has been done in developing regions, including drylands (Lal, 2002b). In order to realize the great potential for CS offered in drylands, efforts are needed to identify soil-specific practices that restore SOC in degraded soils. This reviews specific case studies in different representative agrosystems in drylands, where the potential for carbon storage is assessed under different land-use and landmanagement practices involving irrigation and biofertilizers.

Models for Analysing Tropical Dryland Agricultural Systems

RothC (RothC-26 3) (Coleman and Jenkinson, 1995; Jenkinson and Rayner, 1977) and CENTURY 4.0 (Parton *et al.*, 1987; Parton, Stewardt and Cole, 1988) are the most extensively used SOC simulation models. They have been tested against a variety of long-term agricultural field trials and have also been used in a variety of climate zones, including dryland regions. Both have also been adopted for use in major carbonassessment projects. The two models vary in their complexity. RothC requires fewer data inputs and so is easier to parametize. However, it only deals with soil processes and, consequently, plant residue C is a required input. The CENTURY agro-ecosystem model has similar SOC pools to RothC but has the advantage of additional submodels. Although it is able to handle a great many more land-management options than RothC, it requires an increased array of input variables that require parametization. This is important because the ability of any model to predict accurately depends on the accuracy and trustworthiness of the data used to parametize it.

Approach Adopted for Parametizing Rothc and Century

Data for parametizing the models can be divided into three areas: climate, soil and land management. Studies that contain data that are sufficiently detailed for modelling are few, particularly in dryland regions. Where investigations have been undertaken and detailed information collected, the primary data that are vital for modelling are often not readily available.

Climate parameters may be referred to by investigators reporting the effects of land management on soils, but complete data sets are rarely given. However, climate data are available independently, as from FAOCLIM 2 (2000), which contains a database of more than 28 000 weather stations worldwide. The literature contains the results from many investigations analysing soil properties and the effect of various treatments and practices upon soils. However, there are far fewer studies that combine soil analysis with examination of land management, especially over the longer term. In particular, there has been little research on dryland systems.

For the case studies examined here, CENTURY was parametized and run to equilibrium for between 2 000 and 5 000 years. Scenarios reflecting the recent past and present were then applied. Although CENTURY can run many varied cropping practices, it can only handle one crop at a time. Consequently, intercropping, which is commonly practised in dryland farming systems, could not be incorporated into the scenarios.

RothC was run to equilibrium using the current soil carbon status after initially being run in reverse mode to calculate the necessary plant carbon inputs. In order to model some of the future scenarios for analysing the effects of land management on soil C, the appropriate plant residue inputs were obtained from the CENTURY plant submodel and then used to parametize the RothC land-management files. The results in RothC frequently predict higher levels of soil C than CENTURY. This has been noted before and has been attributed to the fact that SOC tends to turnover faster in CENTURY than in RothC, and, consequently, that RothC requires lower carbon inputs to maintain the same organic carbon content (Falloon and Smith, 2002). The CENTURY plant submodel is quite basic. Therefore, where accurate estimations of plant production and crop yield are required, then alternative models, such as the fully mechanistic plant productivity model WIMOVAC, should be used (Humphries and Long, 1995).

Choice of Systems and Sources of Data

Data from four distinctly different dryland systems in Argentina, India, Kenya and Nigeria were used to model changes in soil C with a variety of farm practices and technologies. These systems had different precultivation stocks of soil C, and had lost different amounts during cultivation. Additions of organic matter to the soil through use of FYM, green manures, legumes in rotations, vermicompost, or use of fallows in rotations, all increased soil C and increased agricultural yields. Trees as part of agroforestry systems further increase soil carbon stocks. Inorganic fertilizer used alone to increase nutrient supply for crops results in declines in soil C in all systems, or only small increases if used with no-tillage. No-tillage increases soil C, although again the accumulation is greatest where organic matter is added to the soil.

The scenarios show that CS in tropical drylands soils can be achieved at the different sites. The land-management practices have been chosen to be in accordance with the current farming systems. Thus, for example, application rates of organic matter are commensurate with quantities that should be available to local farmers. However, at the field level, important trade-offs may occur, preventing adoption of the best strategies for CS. Crop residues may be required for livestock feed or fuel rather than be returned to the fields, or may be sold in difficult times. Animal manures may be burnt for fuel. Many socio-economic factors will interact to determine which scenario or combination of scenarios is implemented in each growing season.

Some of the results predict that soil C can be restored to precultivation levels, or in certain circumstances to above them. The true "native soil carbon level" is often difficult to establish in those systems where agricultural activity has been present for several centuries or millennia such as in the Nigeria and Kenya cases. To achieve quantities of soil C in excess of the "natural level" implies that the agriculture system has a greater productivity than the native system, assuming that C is not being imported. The scenarios that predict the highest CS rates are often associated with the introduction of trees to the system. The inputs of C from trees are more resistant to decomposition than those from herbaceous crops and, consequently, can cause marked increases in the level of soil C (Falloon and Smith, 2002).

Table: *Summary of findings on carbon stocks and rates of accumulation and/or loss in four dryland agrosystems*

	Nigeria	*India*	*Kenya*	*Argentina*
Stocks of soil C before cultivation (tonnes/ha)	8-23	15-20	33-41	50-70
Current stocks of soil C after cultivation (tonnes/ha)	6-12	13-22	18-28	37-41
Effect of conventional tillage practices on soil C (tonnes/ha/year)	- 0.05 to - 0.01	- 0.07 to +0.06	- 0.3 to - 0.1	- 0.17 to - 0.19
Effect of FYM, organic additions, retained plant residues and fallows in rotations (tonnes/ha/year)	+ 0.1 to + 0.3	+ 0.2 to + 0.4	+ 0.4 to + 0.9	-
Effect of trees (tonnes/ha/year)	+ additional 0.05 to 0.15	+ additional 0.5 to 0.7	-	-
Effect of using inorganic fertilizers as sole source of nutrients on soil C (tonnes/ha/year)	- 0.12 to + 0.08	- 0.01	- 0.3	-
Effects of no-tillage (NT)				
NT alone				+ 0.02 +0.1 to 0.25
NT + green manures or FYM				+0.04
NT + inorganic fertilizers				

[1]Effects of conventional tillage are averages for the last 100 years for each site except Kenya where rates are calculated from when each settlement commenced (30 - 50 years).

Case Study 1 - Nigeria - Kano Region

Nigeria includes some of the most densely inhabited areas of semi-arid West Africa. This is not a recent phenomenon as indications of human activity can be traced back well over a thousand years. Consequently, the soils of this region have been subjected to long periods of cultivation. However, in the last 40 years the cultivated area of northern Nigeria has increased from 11 percent to 34 percent of the total land area (Harris, 2000). In particular, the level of agricultural intensity in the Kano Closed - Settled Zone (CSZ) is among the highest in semi-arid West Africa.

Investigations into the farming systems of this region and their effect on soil fertility have been conducted by the organization Drylands

Research, the United Kingdom. Its studies have found a close correlation between the intensity of farming and the adoption of soil-fertility management techniques. Plant production in the Sahel is generally limited by rainfall or nutrients (Breman and De Wit, 1983). However, the economy and infrastructure of northern Nigeria is unsuited to high external inputs such as fertilizers. Consequently, the smallholder farming units operate as low-input systems, with cattle manure being the most common form of organic input.

Physical Aspects

This semi-arid region has a rainy season from May - July until September. However, rainfall is erratic (Mortimore, 2000) and makes crop cultivation particularly difficult. The soils are principally ferruginous tropical soils that are sandy, with poor waterholding capacity and low levels of nutrients and organic matter (Harris, 2000). Nutrient balances can vary between years as crop growth fluctuates with rainfall. The natural vegetation is open forest savannah with a trend towards increasing open grassland where rainfall is lowest.

Farming Systems

Northern Nigeria has been classified into three categories of farming system: intensive, less intensive and extensive (Mortimore, 1989). The intensive systems use permanent annual or biannual cultivation with a cropping intensity of more than 60 percent. The low-intensity systems operate a shrub/short-bush - fallow regime and the cropping intensity is 30 - 60 percent. The extensive system is one of long-bush - fallow and uncultivated areas, where the cropping intensity is typically less than 30 percent. The major crops grown are millet, sorghum, groundnut, sesame and cowpea. As the systems are low input, grain yields are about 1 tonne/ha. Ridging ploughs are primarily used for cultivation and hoeing is commonplace. Applications of manure have to be made carefully to avoid "burning" the crops, and additions are 1 - 7 tonnes/ha. Crop residues are collected for fodder or left to be grazed in the fields. Fallowing is practised in the less intensive systems although the land is still "harvested" through grazing and the collecting of wood and other products. Night-parking (or folding) of livestock is the most effective method of providing manure. Legumes such as cowpea are grown, primarily in higher-intensity systems, to provide nitrogen inputs. Inorganic fertilizers are scarce and seldom available at the optimal time.

Study Sites

The study sites cover a range of population density and include the three categories of farming intensity. The soils and crops grown are similar across all categories (Harris, 2000). Initially, CENTURY was parametized with a natural system of grass, woodland and grazing.

A medium-intensity fire was scheduled for every 10 years and a high-intensity fire event occurred every 30 years.

One thousand years ago, slash and burn episodes were commenced and two seasons of millet cultivation incorporated, repeated every 60 years. In the nineteenth century, the frequency of cultivation events was increased to a 30-year cycle, and then once every 15 years by the commencement of the twentieth century.

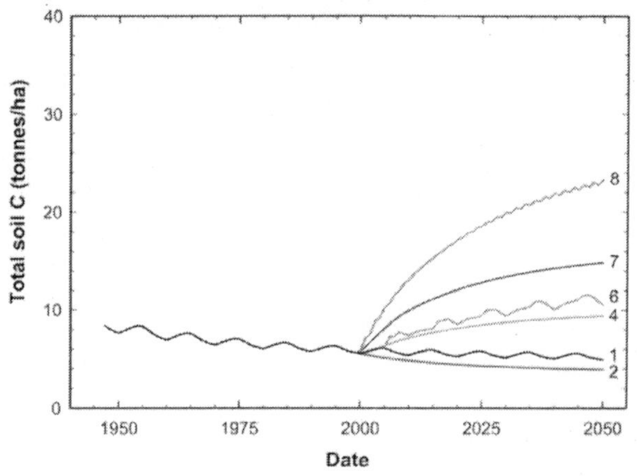

Figure: Total soil carbon for Futchimiram settlement (CENTURY)

Futchimiram, Borno State

This is a low-intensity, or extensive, agropastoral system practising shifting cultivation. Some land has now become degraded. CENTURY was run for the last 60 years with alternate five-year cycles of grazing and millet cropping. Crop residues are grazed and there are no other inputs. This current practice produces a gradual and persistent decline in soil C. The estimated level is slightly above the measured value of soil C for cultivated soils that range between 3.5 and 4.4 tonnes/ha. RothC also predicts that the current practice will decrease soil C slightly over the next 50 years

***Table:** Total soil carbon for Futchimiram settlement*

Scenario[1]	*CENTURY*			*RothC*		
	2000	*2050*	*% change*	*2000*	*2050*	*% change*
	(tonnes/ha)			*(tonnes/ha)*		
1	5.54	4.94	-10.8	5.38	5.18	-3.7
2		3.93	-29.1		3.72	-30.9
5		9.40	69.7		9.70	80.3
6		10.57	90.8		12.45	131.4

Source: CENTURY and RothC.

Table: *Scenarios for modelling land management practices, Futchimiram settlement.*

Scenario	***Land management***
1	Current practice
2	Continuous cultivation
3	Continuous cultivation, no grazing of residues, harvest only grain
4	Inorganic fertilizer only (100 kg/ha urea), no grazing
5	Plant residues average 0.5 tonnes/ha/year, no grazing of residues
6	5-year fallow, 5-year cultivation, 2 applications FYM 3 tonnes/ha, graze residues
7	Continuous cultivation, FYM 1.5 tonnes/ha/year, graze residues
8	Continuos cultivation, FYM 1.5 tonnes/ha/year
	Plant residues 0.5 tonnes/ha/year, no grazing

Effect of Loss of Fallow

Removing the fallow period only (Scenario 2) results in a greater decline in soil C over subsequent years, with both models predicting similar reductions. Preventing grazing of the crop residues under these conditions (Scenario 3) has very little effect on soil C.

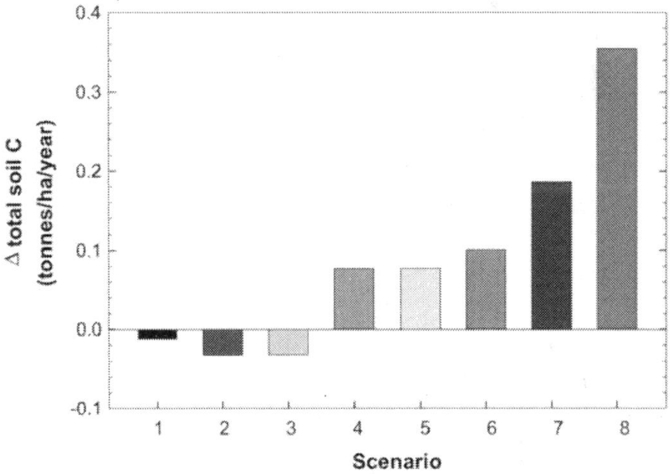

Figure: *Average annual change in total soil carbon for Futchimiram settlement (CENTURY)*

Effect of Organic Inputs

The addition of FYM (two applications of 3 tonnes/ha in each five-year cropping cycle, average 0.6 tonnes/ha/year) has a positive effect on soil C, the two models indicating an increase of 5 - 7 tonnes C/ha over the next 50 years (0.08 tonnes C/ha/year). The outcome of further scenarios using different combinations of fallow and organic inputs. The gradual increase in soil C occurs with FYM, retaining plant residues and no grazing after harvest. In this case, soil stocks rise from 6 to 24 tonnes C/ha over 50 years.

Effect of Inorganic Fertilizer

The use of inorganic fertilizer with no other organic inputs, and no grazing or fallow (Scenario 4), leads to a modest increase in CS (0.08 tonnes C/ha/year).

Summary

This system will not yield a positive soil carbon balance without increased organic inputs. Although inorganic fertilizers can lead to CS, they bring with them a carbon cost that will result in an overall negative balance of the complete carbon budget.

Kaska, Yobe State

This is a low-intensity agropastoral farming system covering lowland, upland and some wetland areas. Only the lowland soils are modelled here. There is intercropping of legumes and grains, and crop residues are fed to livestock. Manure application to fields is low and long bush fallows are used. CENTURY was parametized for the last 50 years with a sevenyear cycle of 4 years of grazing and 3 years of millet - cowpea - millet cropping. FYM (0.75 tonnes/ha) was added in the first year of each cropping cycle. Cultivated soils have carbon contents of 4.5 - 7.0 tonnes/ha. The CENTURY model calculates a current carbon content of 7.7 tonnes/ha, and RothC gives a similar result. Both models suggest that with current practice the system is close to steady state.

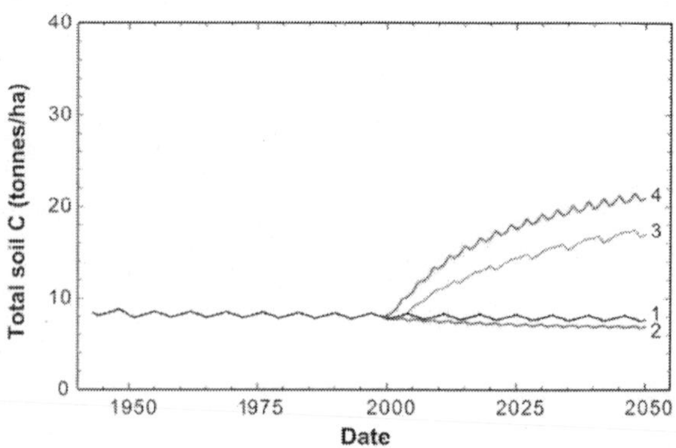

Figure: Total soil carbon for Kaska settlement (CENTURY)

Effects of Fallows and Organic Inputs

Removing the fallow from the current practice leads to a slight decline in soil C in subsequent years. Applying 3 tonnes/ha FYM to each millet crop (average 1.3 tonnes/ha/year over seven-year cycle) will produce a marked increase in soil C representing a CS rate of 0.18 tonnes/ha/year. This increase will be further enhanced if the fallow is removed, because the manure application rate will now average 2 tonnes/ha/year.

Summary

The scenarios for Kaska illustrate the effect of fallow periods on stocks of soil C. Where the cropping regime is adding very little organic matter to the soil, fallows will often have a positive effect if correctly managed.

However, if the cropping practice is accumulating significant organic matter in the soil, any interruption, such as fallowing, will decrease the overall potential for CS.

Table: *Total soil carbon for Kaska settlement*

Scenario[1]	*CENTURY*			*RothC*		
	2000	*2050*	*% change*	*2000*	*2050*	*% change*
	(tonnes/ha)			*(tonnes/ha)*		
1	7.73	7.64	-1.2	7.33	7.87	7.4
3		16.94	119.1		15.57	112.4

Source: CENTURY and RothC.

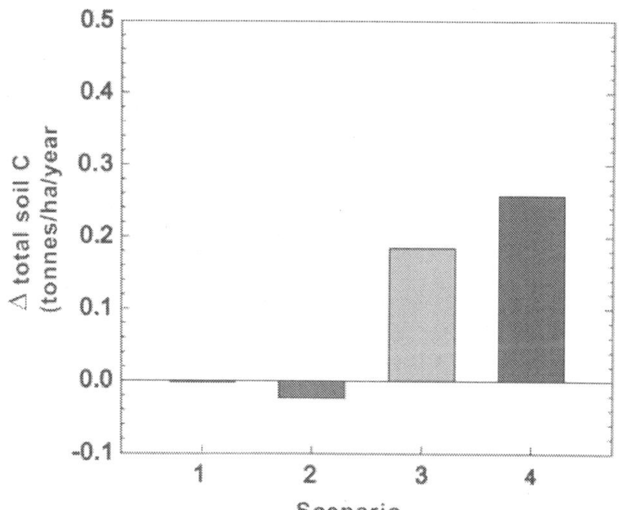

Figure: *Average annual change in total soil carbon for Kaska settlement (CENTURY).*

Table: *Scenarios for modelling land management practices, Kaska settlement.*

Scenario	***Land management***
1	Current practice
2	Continuous cultivation, millet-cowpea
3	Cultivation-fallow, FYM 3 tonnes/ha, to millet
4	Continuous cultivation, FYM 3 tonnes/ha, to millet

Dagaceri, Jigawa State

This is a region undergoing rapid intensification. It is an agropastoral system with shrub or short-bush fallowing. The length of fallow has decreased as the area of arable land has expanded. Both legumes and grains are grown, but land degradation is a problem. As the fallow period shortens, farmers rely increasingly upon manures to maintain soil fertility. The parametization for CENTURY was initially the same as for Kaska but then, 30 years ago, cropping was increased to five years out of seven. Millet and cowpea are cropped alternately with 1.5 tonnes/ha manure added to each millet crop (average 0.64 tonnes/ha over a seven-year cycle). The additional manure input associated with the increase in cropping intensity results in an increase in soil C, predicted by both models, and this rise continues in subsequent years. Field measurements vary according to previous cropping intensity and range from 3 to 17 tonnes C/ha.

Table: *Total soil carbon for Dagaceri settlement (CENTURY and RothC)*

Scenario[1]	*CENTURY*			*RothC*		
	2000 (tonnes/ha)	*2050* (tonnes/ha)	*% change*	*2000* (tonnes/ha)	*2050* (tonnes/ha)	*% change*
1	12.43	15	20.7	14.69	20.43	39.1
5		22.22	78.8		29.55	101.2

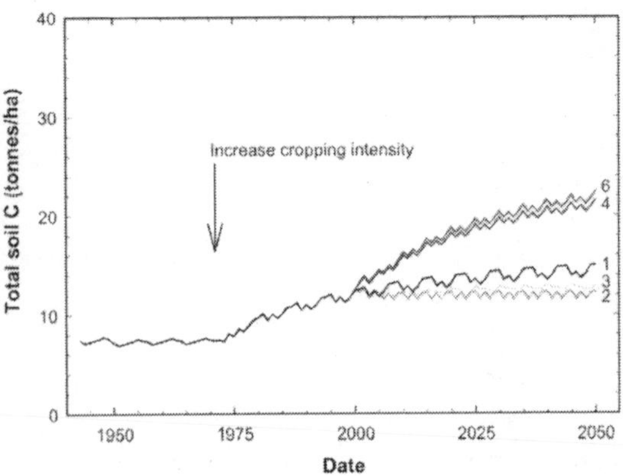

Figure: *Total soil carbon for Dagaceri settlement (CENTURY)*

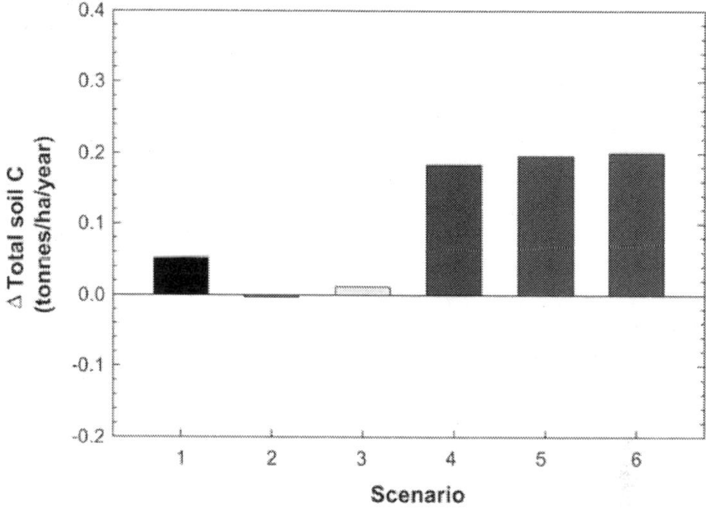

***Figure:** Average annual change in total soil carbon for Dagaceri settlement (CENTURY)*

Effects of Fallows and Organic Inputs

The manure input to each millet crop is increased to 3 tonnes/ha in all scenarios (2 - 6). Carbon sequestration rates of 0.18 - 0.20 tonnes/ha can be achieved, with slight differences depending on whether the fallow is retained and crop residues grazed or ungrazed. However, if fallow elimination is accompanied by harvesting all aboveground material (Scenario 3), CS is virtually halted even if manure application is maintained.

The complete removal of trees from the system results in a net loss of soil C in spite of the increased application of manure.

Summary

This system shows that SOM can be maintained and increased even with intensification of cropping, provided there are legumes in the rotation. However, careful management of crop residues is vital as is the preservation of trees.

iv. Tumbau, Kano Closed - Settlement Zone

This is a highly intensive agricultural area. All the land is cultivated although degradation reportedly affects less than 10 percent of the land area. There is a highly integrated crop and livestock production system with intercropping of legumes, intensive manuring and inorganic fertilizer application. There is virtually no grazing land so animals have to feed on crop residues and fodder from nearby fields.

CENTURY was run for the last 50 years with a millet - cowpea rotation, 6 tonnes/ha of manure applied to the millet (average 3 tonnes/ ha/year) and all aboveground plant material harvested. This system is now close to steady state with soil C stocks at 9.8 tonnes C/ha. This compares with an average of 10.5 ± 1.7 tonnes C/ha measured at cultivated sites. RothC calculates a value of 11.3 tonnes/ha for 2000 but predicts that a higher soil carbon level will be reached by 2050.

Table: *Scenarios for modelling land management practices, Dagaceri settlement.*

Scenario	***Land Management***
1	Current practice
2	Remove trees
3	No grazing of residues, harvest all aboveground
4	Continous cultivation, millet - cowpea
5	FYM average 1.29 tonnes/ha/year,fallow, graze residues, harvest only grain
6	No grazing of residues, harvest only grain

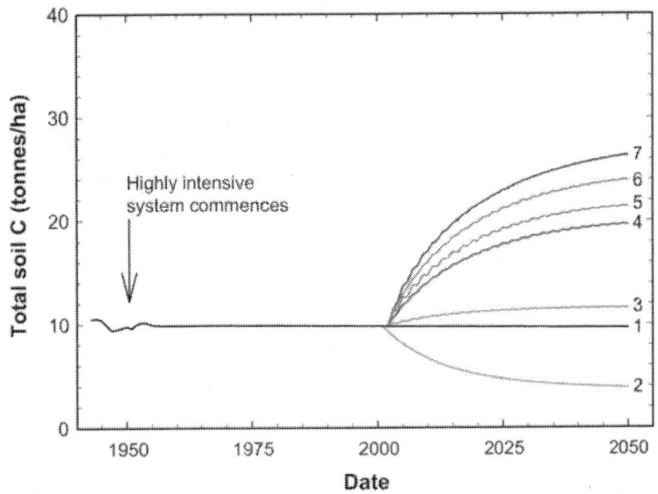

Figure: *Total soil carbon for Tumbau settlement (CENTURY)*

Effects of Organic Inputs

Additional FYM inputs have a marked impact on soil C, especially when they approach the maximum (7 tonnes/ha) normally applied in this region, i.e. an annual input of 6.75 tonnes/ha will result in the sequestration of 0.20 tonnes C/ha/year over the next 50 years. Although adding Nfixing trees and plant residues to the fields will further increase soil C, the requirement of the former for livestock feed may exceed the capacity of the current farming system.

Effect of Inorganic Fertilizer

Replacing the manure input to this system with inorganic fertilizer (urea 100 kg/ha, Scenario 2) results in a large reduction in soil C, with stocks falling by more than 0.1 tonnes C/ha/year.

Table: *Total soil carbon for Tumbau settlement (CENTURY and RothC)*

*Scenario**	*CENTURY*			*RothC*		
	2000	***2050***	*% change*	***2000***	***2050***	*% change*
	(tonnes/ha)			*(tonnes/ha)*		
1	9.82	9.64	-1.8	11.3	13.39	18.5
3		11.54	17.5		14.96	32.4
6		23.9	143.4		31.7	180.5

Table: *Scenarios for modelling land management practices, Tumbau settlement.*

Scenario	*Land management*
1	Current practice
2	Inorganic fertilizer (110 kg/ha urea)
3	FYM 3.75 tonnes/ha/year
4	FYM 6.75 tonnes/ha/year
5	FYM 3.75 tonnes/ha/year, add nitrogen-fixing trees
6	FYM 6.75 tonnes/ha/year, plant residues 2 tonnes/ha
7	FYM 6.75 tonnes/ha/year, harvest only grain

Summary

Provided that adequate organic matter is returned to the soil, these intensive-farming systems should maintain soil C and there is also scope for CS. These results are in agreement with findings in the field that provide no evidence for a decline in SOM in spite of increased cultivation pressure.

However, the ability to realize future CS will depend on a careful balance between cropping and livestock husbandry and the overall capacity of the system. Maintaining crop yields through the application of inorganic fertilizer alone will probably result in substantial losses of SOM.

Conclusions From Northern Nigeria Cases

The modelling of farm data for the drylands of northern Nigeria shows that soil carbon stocks can be increased from a low base with a variety of technologies and practices already available to farmers.

The total amounts of C that can be sequestered with the use of legumes, fallow periods, FYM and retention of plant residues varies between 0.1 and 0.3 tonnes C/ha/ year. This amount rises when trees are also cultivated.

Soil C is lost when only inorganic fertilizers are used to maintain soil fertility - some 0.1 tonnes C/ha/ year in the intense systems of the Kano CSZ. Continuous cultivation results in small year-on-year losses of C where there are no additional inputs of organic matter. In spite of considerable intensification of current systems (shortening of fallow periods), farmers are maintaining carbon stocks. The benefits of maintaining trees in the landscape are shown in these modelled scenarios.

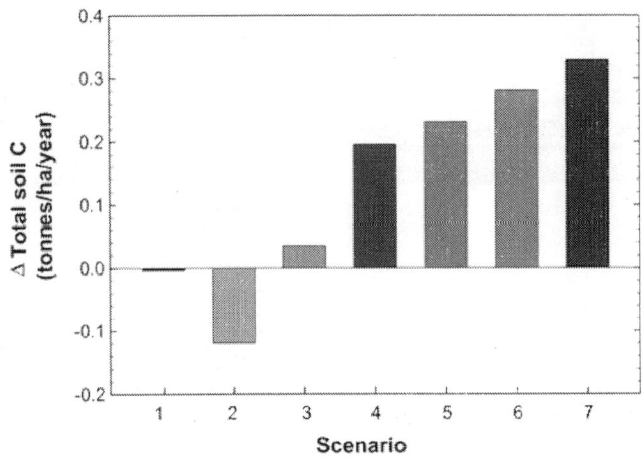

Figure: *Average annual change in total soil carbon for Tumbau settlement (CENTURY)*

Case Study 2 - India - Andhra Pradesh and Karnataka States

More than half of India's farmers of India live in climate regions that can be described as semi-arid. Recent decades have seen increases in crop yields that have been attributed to the green revolution. However, the associated technologies, e.g. irrigation and inorganic fertilizer are expensive, and not readily accessible to the rural poor. These practices may lead to declines in soil fertility, and are also dependent upon fossil-fuel energy (Butterworth, Adolph and Satheesh, 2002). In fact, nearly two-thirds of the arable land in India remains dependent solely upon rainfall for agricultural production. The Natural Resources Institute (the United Kingdom), the Deccan Development Society and the BAIF Institute for Rural Development (India) have studied soil fertility management in Medak District in Andhra Pradesh and in Tumkur District in Karnataka. They have shown that there is an increasing awareness about technologies for maintaining and improving soil fertility, identifying at least 14 different practices from legume cultivation to vermicompost production. Most involve maintaining and increasing the organic matter content of the soil. These studies provide an opportunity to investigate the effect that these soil-fertility improvement techniques may have on CS.

Physical Aspects

Medak District forms part of the tableland of the Deccan Plateau that extends from Andhra Pradesh into Karnataka. The climate consists of a mild winter period (rabi, November - February), a hot and dry summer (March - May) and the southwest monsoon, when more than 80 percent of rainfall occurs (kharif, June - October). Average rainfall is slightly less than 900 mm. The hottest month is May, just before the onset of rains, when the maximum daytime temperature can reach 40 °C. Conversely, night temperatures can drop to 6 °C in December. Moisture availability for crop growth ranges from 120 to 150 d.

The major soil types are Alfisols and Vertisols. The former include red lateritic soils comprising loamy sands, sand loams and sandy clay loams and are usually nonsaline. The Vertisols, black cotton soils, are potentially more productive with a higher waterholding capacity, moderately alkaline and with a highly soluble salt content. They comprise clay loams, clays and silty clays. The organic carbon content of soils in the area is usually 0.5 - 1 percent. The land lies between 500 - 600 m above sea level, and very little natural vegetation remains. The tropical dry deciduous forest has mainly been felled except on protected government land.

Farming Systems

The farming systems in this region have a high degree of integration between livestock, crops and trees (Pound, 2000). There is a huge agrobiodiversity that is combined with off-farm activities, particularly in Medak District. This situation makes for an efficient use of the limited land resources and acts as insurance against unpredictable weather conditions, a frequent problem in drylands. The average farm size is 2.6 ha (Butterworth, Adolph and Satheesh, 2002). The small and marginal farmers can grow at least eight varieties of crops per half hectare. The cattle population has been falling continuously since the 1980s. This has important implications for agriculture, not only because animal wastes are an important source of organic matter for soils, but also because bullocks make a vital contribution as draught animals, not least in the transport of FYM. Tillage is commonly performed with very basic implements. The greatest shift towards mechanized cultivation has been in Karnataka.

The predominant crops grown in Medak District are paddy, sorghum and maize while irrigated sugar cane is an important cash crop. The major crops grown in Karnataka are paddy, sorghum, finger millet, pearl millet, pigeon pea, green gram, and groundnut (Reddy, 2001).

Farmers attach a high priority to maintaining soil fertility while inorganic fertilizers are ranked poorly in terms of maintaining soil quality. However, many farmers use them because organic alternatives are often unavailable and because the inorganic fertilizers are subsidized. The importance attached to soil fertility can have unforeseen consequences such as the sale of FYM by the poorest farmers to their more affluent counterparts. The ultimate effect of this will be an increase in the fertility and carbon content of some soils while degrading neighbouring areas.

Modelling Soil Carbon in The Study Villages

The CENTURY agro-ecosystem model was parametized using climate and soil data and run to equilibrium commencing with a grass/woodland system to represent the natural vegetation of this region. Low-intensity grazing and fires occurring every 30 years were included in the cycle.

One thousand years ago, the effects of human interference were introduced with slash and burn events together with the cultivation of a grain crop. The frequency of these events was increased slowly and by the start of the twentieth century, cultivation periods of four years out of ten were introduced.

In the mid-twentieth century, one year of grazed fallow was followed by four years of cropping (sorghum - kharif, and cowpea - rabi). Cultivation consisted of hand or bullock ploughing and hand hoeing. Average annual applications of FYM were 2.1 tonnes/ha over the five-year cycle and all aboveground material was harvested. For the final 30 years of the twentieth century, cultivation was adjusted to reflect current practices. The RothC model was run to equilibrium in the mid-twentieth century, and then parametized using current practice and plant residue quantities calculated by CENTURY.

Data on five types of farm and location are presented below:

i. large mixed dryland farm (Lingampally);
ii. small dryland farm, no livestock (Lingampally);

iii. large farm using irrigation (Yedakulapaly);

iv. small mixed drylands farm (Metalkunta);

v. small mixed drylands farm (Malligere);

Analysis of Land Management by a Large Mixed Farm, Lingampally Village, Medak District

This is a holding of slightly more than 5 ha on predominantly Vertisols. Livestock are fed fodder and plant residues from the fields. Consequently, no plant material is returned to the soil. The animals are also grazed on local common land. The livestock provide manure (2 - 3 tonnes/ha/year). In addition, inorganic fertilizer has been used in recent years (30 - 45 kg/ha of di-ammonium phosphate). Crops modelled are sorghum (kharif May - September) and cowpea (rabi October - January). Century predicts that the current farming practice is resulting in a nearly stable soil carbon content of about 19 tonnes/ha declining to 2050 by nearly 1.5 tonnes/ha. RothC shows a lower total soil carbon content of 16 tonnes/ha in 2000, declining to 15 tonnes/ha by 2050.

Effect of Inorganic Fertilizer

If the application of inorganic fertilizer were halted and the quantity of FYM increased to replace the nitrogen input and maintain yields (increase manure application by 0.6 tonnes/ha, Scenario 3), soil C would increase by only 0.85 tonnes/ ha by 2050 and there would be no net gain in soil C.

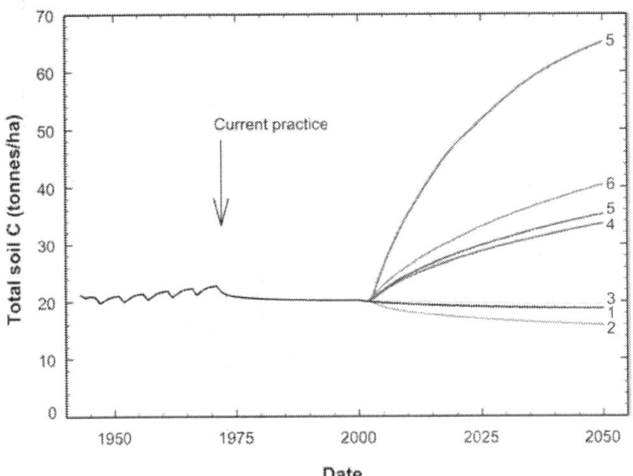

Figure: *Total soil carbon for a large farm (5 ha), Lingampally village (CENTURY).*

Conversely, replacing all organic additions with inorganic fertilizer would result in a decrease in soil C of more than 4 tonnes/ha by 2050, a loss of 0.09 tonnes/ha/ year (Scenario 2). Thus, inorganic fertilizer results in a fall in soil C.

Effect of Farmyard Manure

Doubling the current annual input of FYM to 4 - 6 tonnes/ha without applying inorganic fertilizer has a marked effect on soil C: 0.27 tonnes/ha/year is sequestered in the next 50 years, with the system still not at a steady state. Similarly, RothC shows a continuing rise, reaching 28.3 tonnes/ha by the same date.

Table: *Total soil carbon for a large farm Lingampally village*

Scenario[1]	*CENTURY*			*RothC*		
	2000	*2050*	*% change*	*2000*	*2050*	*% change*
	(tonnes/ha)			*(tonnes/ha)*		
1	20.2	18.8	-7.2	18.3	17.9	-2.3
4		33.5	65.9		28.3	54.2
6		40.4	100		31.0	68.8

Source: CENTURY and RothC.

Effect of Other Organic Inputs

Adding only modest amounts of vermicompost (100 kg/ha) and green manure (250 kg/ha) in addition to the doubled FYM (Scenario 5) has a limited effect, while 2 tonnes/ha plant residues makes a bigger contribution, 0.4 tonnes/ha/year.

Effect of Trees

This farm has a capacity for introducing N-fixing trees such as *Glyricidia,* which creates Scenario 7 when added to the manure and plant residues.

Table: *Scenarios for modelling land management practices, large farm Lingampally village.*

Scenario	*Land management*
1	Current practice
2	FYM 3 tonnes/ha/year
3	FYM 3 tonnes/ha/year, green manure 500 kg/ha/year, vermicompost 250 kg/ha/year
4	As current practice but incorporate crop residues into soil
5	FYM 3 tonnes/ha/year, leave plant residues
6	FYM 3 tonnes/ha/year, plant residues, green manure, vermicompost
7	FYM 6 tonnes/ha/year, plant residues, green manure, vermicompost

After 10 years, the trees are cut annually for wood. The result is a very large increase in soil CS of 0.4 tonnes/ha/year. This increase exceeds that which would be obtained by increasing the manure application by four times.

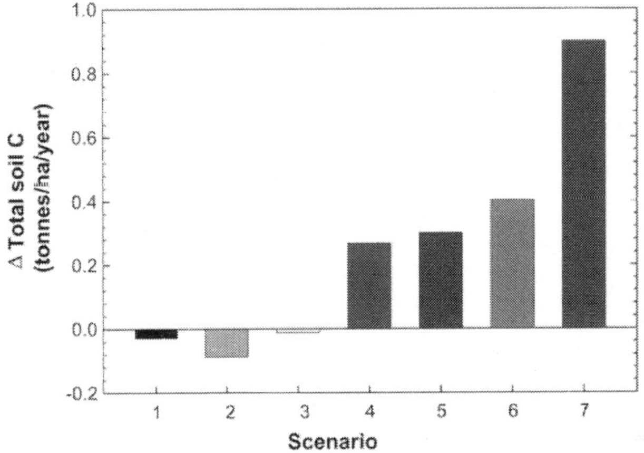

Figure: *Average annual change (over 50 years) in total soil carbon for a large mixed farm, Lingampally village (CENTURY)*

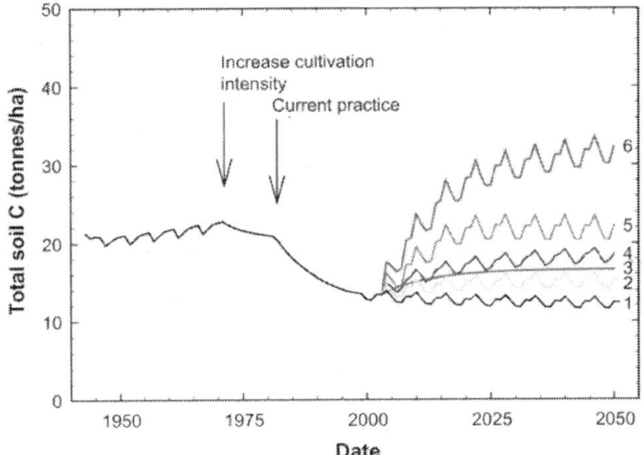

Figure: *Total soil carbon for a small rainfed farm, Lingampally village (CENTURY).*

Summary

The two soil carbon models are in fairly close agreement for this farming system. The current practice is nearly sustainable with only a small decline (< 2 percent) in soil C predicted over the next 50 years. However, cattle are currently being grazed on other land and, consequently, some C is effectively being mined from elsewhere.

A modest increase in organic material would be required to replace the inorganic fertilizers currently used. Further organic inputs could increase soil CS substantially. Introducing trees is likely to have a marked effect on soil C and would simultaneously increase aboveground carbon storage. However, a greater proportion of trees may have been introduced to the model than would be feasible and, therefore, the degree of CS that is practicable for this farming system may be overestimated.

Analysis of Land Management by a Small Rainfed Farm, Lingampally Village, Medak District

This holding is less than 1 ha and the Vertisol has become very degraded. Animals were kept in the past but have been sold and currently sorghum is the only major crop grown. All aboveground material is harvested and removed. Some FYM is applied at different times, equivalent to 3.9 tonnes/ha/year. The CENTURY model shows how current practices have produced a marked decline in soil C, reaching some 13 tonnes/ha in 2000. However the system is reaching steady state and the current land-management practice is not predicted to cause much further decline in soil C in the next 50 years. RothC yields a slightly higher carbon content in 2000 but predicts a slightly greater decline (1.8 tonnes/ha) by 2050.

Effect of Retained Residues

Harvesting only the grain and the returning crop residues to the soil would produce a positive carbon balance, sequestering 0.05 tonnes/ha/year.

Effect of Farmyard Manure

Increasing the input of FYM by 50 percent to 6 tonnes/ha/year would increase soil C by 3.8 tonnes/ ha by 2050. RothC predicts a rise of 3.0 tonnes/ha.

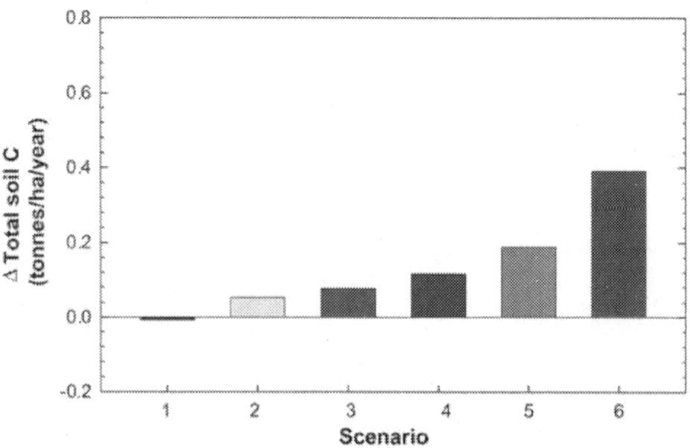

Figure: Average annual change in total soil carbon for a small rainfed farm, Lingampally village (CENTURY)

Effect of Legume Crops

Adding cowpea the rotation would sequester 0.12 tonnes/ha/year. Introducing Nfixing trees (cut every 2 years after an initial 10 years) to the sorghum-only system with FYM at 4 tonnes/ha, would yield a sequestration rate of 0.19 tonnes/ha/year. Combining all treatments would lead to a soil CS rate of 0.39 tonnes C/ha/year.

Table: *Total soil carbon for a small rainfed farm, Lingampally village.*

*Scenario**	*CENTURY*			*RothC*		
	2000	*2050*	*% change*	*2000*	*2050*	*% change*
	(tonnes/ha)			*(tonnes/ha)*		
1	12.77	12.41	-2.8	16.07	14.92	-7.2
3		16.58	29.8		19.11	18.9

Source: CENTURY and Roth C.

Table: *Scenarios for modelling land management practices, for a small rainfed farm, Lingampally village*

Scenario	***Land management***
1	Current practice
2	Harvest only grain
3	FYM 6 tonnes/ha/year
4	Add legume (cowpea)
5	Add trees,e.g. *Glyricidia*
6	All additions

Summary

The modelling suggests that the loss of soil C can be reversed for this type of farm, even on degraded soil. In addition to direct organic inputs, this example illustrates the importance of a leguminous crop, and the inclusion of trees such as *Glyricidia.*

Analysis of Land Management for a Large Farm Using Irrigation, Yedakulapaly Village, Medak District

The farm comprises 4.5 ha of irrigated Vertisol. Water is pumped from a borehole. For the purposes of modelling, 5 cm of water is applied when the soil waterholding capacity is < 25 percent during the vegetative growth stage of the crop. Livestock were originally kept but were sold some years ago because there was insufficient labour to collect fodder and tend to the cattle during grazing. Only inorganic fertilizers are now used - 650 kg/ha di-ammonium phosphate and 150 kg/ha of urea annually. Crops are grown in all three seasons (summer - sorghum, kharif - legume, rabi - legume) and all aboveground material is harvested and removed. For the purposes of modelling, cowpea is used for the legume crop.

Century calculates that current practices have depleted soil C to 13.26 tonnes/ha and predicts no further reduction in the next 50 years.

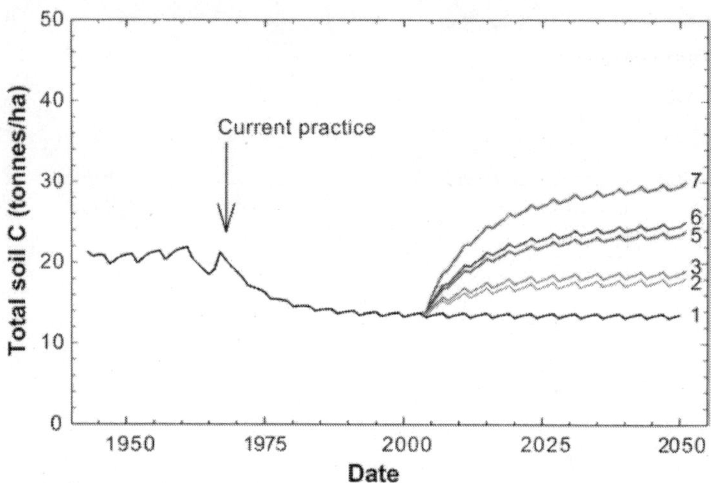

Figure: *Total soil carbon for a large farm using irrigation and cultivating three crops per year, Lingampally village (CENTURY)*

Effect of Organic Inputs

Adding 3 tonnes FYM/ha/year as per Scenario 2 would increase soil C by 4.4 tonnes/ha to 2050, sequestering 0.09 tonnes/ha/year. Including 500 kg/ha/year of green manure and 250 kg/ha/year of vermicompost would increase soil organic C by a further 1 tonne, sequestering 0.11 tonnes/ha/ year.

Table: *Scenarios for modelling land management practices, large farm using irrigation Lingampally village*

Scenario	*Land management*
1	Current practice
2	FYM 3 tonnes/ha/year
3	FYM 3 tonnes/ha/year green manure 500 kg/ha/year, vermicompost 250 kg/ha/year
4	As current practice but incorporate crop residues into soil
5	FYM 3 tonnes/ha/year, leave plant residues
6	FYM 3 tonnes/ha/year, plant residues, green manure, vermicompost
7	FYM 6 tonnes/ha/year, plant residues, green manure, vermicompost

A similar increase in soil C can be obtained with no extra organic additions if only the grain is harvested and the crop residues are returned to the soil. When 3 tonnes FYM/ha/year is applied and only the grain harvested, soil C increases by more than 10 tonnes/ha by 2050, equivalent to 0.20 tonnes/ha/year. Applying all the organic inputs and increasing the FYM component to 6 tonnes/ha would sequester 0.32 tonnes/ha/year.

Summary

A combination of leaving crop residues and adding manure could make a significant contribution to soil C in this kind of farming system, in which current practices have already caused a significant decline in soil carbon stocks. However, inorganic fertilizer and irrigation continue to have a carbon cost, resulting in smaller increases under the various scenarios.

Additions of FYM can offset the nutrients applied in inorganic fertilizers. For example, removing inorganic fertilizer from Scenario 6 maintains yield without affecting soil C. Removing irrigation from the above scenario does not have a detrimental effect on soil C, but nor is there an apparent effect on yield.

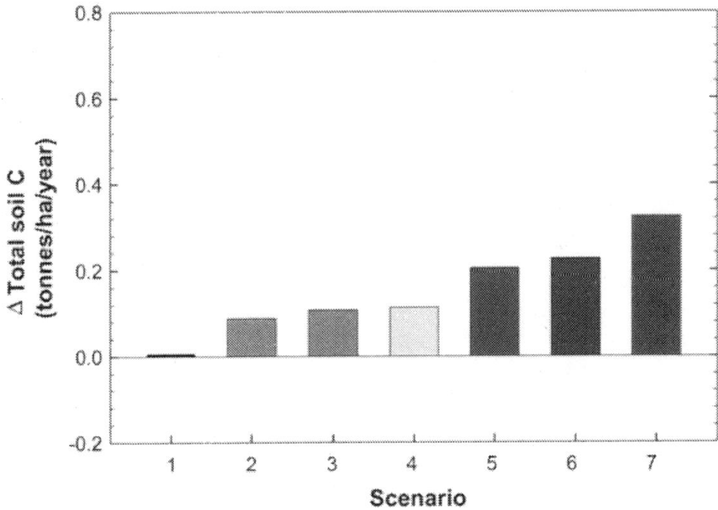

Figure: *Average annual change in total soil carbon for a large farm using irrigation, Lingampally village (CENTURY).*

Analysis of Land Management for a Small Mixed Crop and Livestock Farm, Metalkunta Village, Medak District

This landholding of 2.7 ha has a very large variety of crops on lateritic red soil. Livestock are kept, a good cultivation pattern is practised and there are also many trees present around the farm. All aboveground produce is harvested so that crop residues can be fed to the animals. Cattle are also grazed on nearby land and fodder is brought in. The average yearly manure application is 2 tonnes/ha, and vermicompost is added to the land.

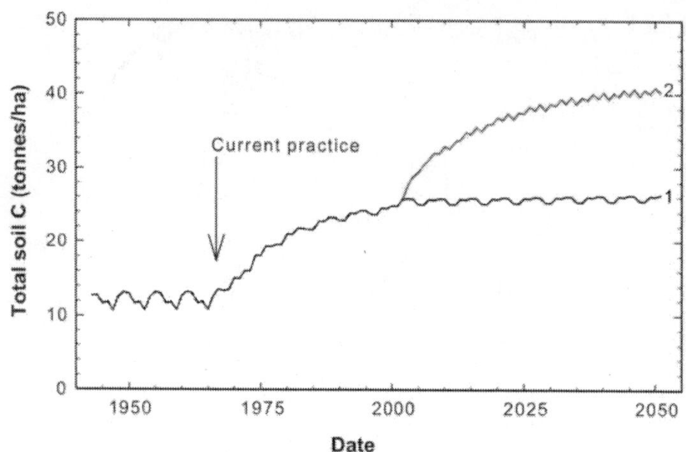

Figure: *Total soil carbon for a small mixed crop and livestock farm, Metalkunta village (CENTURY)*

Century was parametized with a sorghum - cowpea - millet - cowpea - maize - cowpea rotation to reflect the varied cropping pattern. The model was run with this scenario from the late 1960s. The model suggests that this system of land management would have approximately doubled the soil carbon content to nearly 25 tonnes/ha by 2000 and would reach 26 tonnes C/ha by 2050. RothC predicts 25 tonnes C/ha in 2000 rising to 29.1 tonnes/ha in 2050. This mixed cropping and livestock system has a substantial positive impact on soil C.

Table: *Total soil carbon for a small mixed crop and livestock farm, Metalkunta village*

*Scenario**	*CENTURY*			*RothC*		
	2000	*2050*	*% change*	*2000*	*2050*	*% change*
	(tonnes/ha)			*(tonnes/ha)*		
1	24.9	26.1	5.2	25.0	29.2	16.7
2		40.9	64.7		49.2	96.6

Source: CENTURY and RothC.

Table: *Scenarios for modelling land management practices, small mixed crop and livestock farm, Metalkunta village*

Scenario	*Land management*
1	Current practice
2	Leave plant residues

Effects of Retaining Crop Residues

Returning crop residues to the soil would greatly increase the carbon content, sequestering 0.3 tonnes C ha/year in the next 50 years, with the soil carbon content reaching 40.9 tonnes/ ha by 2050. RothC calculates that leaving crop residues would increase soil C to 49.2 tonnes/ha.

Summary

The farmer is currently practising very good land management, and the models calculate that the soil carbon content has already been substantially increased. However, this is being accomplished in part by grazing cattle on other lands and bringing in fodder. This means that C is effectively being mined from elsewhere. Thus, although returning crop residues to the soil makes a further significant contribution to soil C, this would require importing more fodder to substitute for the plant residues, which again implies mining C.

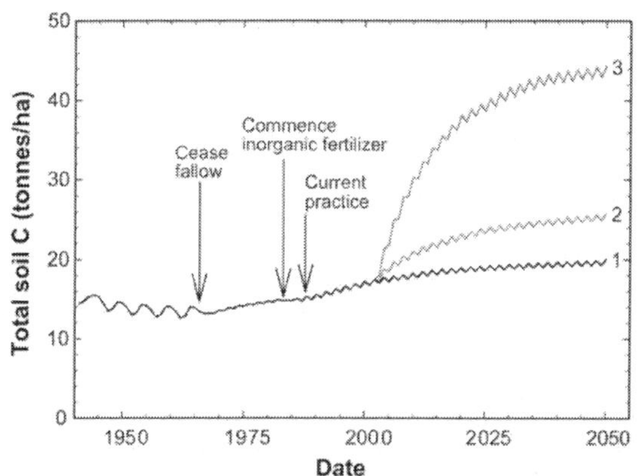

Figure: *Total soil carbon for a small farm, Malligere village, Tumkur District (CENTURY)*

Table: *Total soil carbon for a small farm, Malligere village, Tumkur District.*

*Scenario**	*CENTURY*			*RothC*		
	2000	*2050*	*% change*	*2000*	*2050*	*% change*
	(tonnes/ha)			*(tonnes/ha)*		
1	17.24	19.97	15.8	19.82	24.32	22.7
2		25.73	49.2		34.2	72.6

Source: CENTURY and RothC.

Table: *Scenarios for modelling land management practices, small farm, Malligere village*

Scenario	*Land management*
1	Current practice
2	Replace inorganic fertilizer with FYM
3	Add trees, *Glyricidia*

Analysis of Land Management for a Small Farm, Malligere Village, Tumkur District Karnataka State

This farm covers 2 ha on a red sandy soil. Tractors are hired for transport, livestock are kept and allowed to graze the crop residues, but additional fodder is brought in. The kharif crops modelled are sorghum and millet and the rabi crop is cowpea. On average, 3 tonnes/ ha/ year FYM is applied. Inorganic fertilizers are also used: 75 kg/ha/ year of di-ammonium phosphate and 75 kg/ha/year of urea. Modelling with CENTURY shows that current practices are increasing soil C and that this will continue to rise by a further 2.7 tonnes/ha in the next 50 years, reaching almost 20 tonnes C/ha in 2050. RothC also shows that the current practice is increasing soil C but shows an enhanced effect. Soil C in 2000 is predicted to be 19.8 tonnes/ha, rising to 24.3 tonnes/ha in 2050.

Effect of Organic Additions

Inorganic fertilizer has a carbon cost, so replacing the N supplied by the inorganic fertilizer with manure (annual addition 6.3 tonnes/ ha) enhances the CS rate to 0.17 tonnes/ ha/year. The soil carbon level reaches 25.7 tonnes/ha by 2050. RothC predicts a much bigger effect on SOC of replacing the fertilizer with FYM: soil C is predicted to rise to 34.2 tonnes/ha by 2050.

Effect of Trees

Adding trees such as *Glyricidia* to the system (Scenario 3), which are cropped annually for wood after ten years, makes a very large difference, increasing CS to 0.54 tonnes/ ha/year. Both models show similar trends, with a switch from inorganic to organic fertilizer increasing soil C significantly. RothC predicts a larger increase. Cessation of inorganic fertilizer use would remove the negative carbon balance associated with its production, and the addition of trees may be sufficient to offset the inputs of C that arise from fodder grown outside the farm system. However, if tractors are used to transport fodder, the carbon budget for the system will probably be negative.

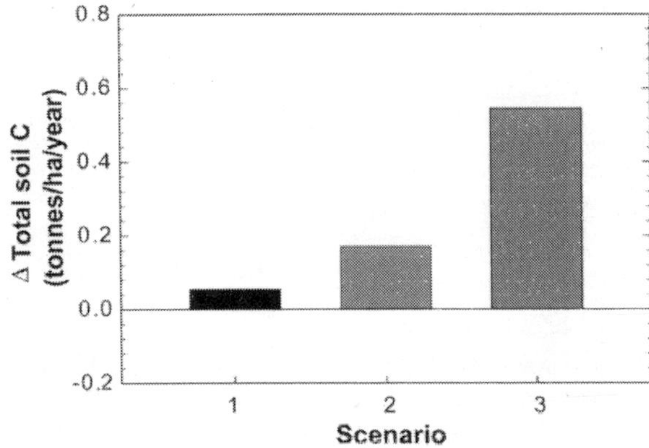

Figure: *Average annual change in total soil carbon for a small farm, Malligere village, Tumkur District (CENTURY)*

Conclusions From India Cases

The modelling of farm data for the drylands of India shows that soil carbon stocks can be increased with a variety of technologies and practices available to farmers.

It also shows that some practices result in substantial declines in carbon stocks, particularly the use of inorganic fertilizer as the sole source of nutrients, and the continuous cultivation of cereals. In the large (5 ha) mixed farm, inorganic fertilizer results in the loss of 0.1 tonnes/ha/year, whereas the use of either FYM, green manures, vermicompost and/or plant residues produces increases of 0.2 - 0.4 tonnes/ha/year. The use of agroforestry substantially increases belowground CS to 0.9 tonnes/ha/year. The models show substantial declines in soil C on the small farms that cultivate only sorghum and those that intensively cultivate three irrigated crops per year - a loss of 5 tonnes C/ha in a 25-year period. However, it appears that these falls can be reversed within 5 - 10 years with the adoption of legumes in rotations, the addition of FYM manure and cultivation of trees.

The small mixed cropping farm, with cereals and cowpea rotations and livestock, increases soil C from a stock of about 13 tonnes C/ha to 24.8 tonnes C/ha in 25 years. It can increase this to more than 40 tonnes C/ha in the next 50 years if plant residues are added to the soil.

For full accounting of the C used or sequestered in these farms, it is important to consider the high energy cost of nitrogen fertilizer manufacture (65.3 MJ/kg for N, 7.2 MJ/kg for phosphorus, 6.4 MJ/kg for potassium) (Pretty *et al.*, 2002), the use of mechanized operations on farms, the cost of irrigation and mechanical transport, and the issue of the transfer of C in feed or livestock themselves from one farm to another or from grazing areas to cropped fields.

There are clear benefits for farmers and soil C if leguminous crops are included in rotations and in agroforestry systems.

Case Study 3 - Kenya - Makueni District

Arid and semi-arid lands occupy about two-thirds of Kenya (FAO, 1999b). The major factors limiting crop growth and production in these areas of Kenya are erratic rainfall, poor husbandry and declining soil fertility caused by continuous cultivation (Kenya Agricultural Research Institute, 1999). Drylands Research has examined the sustainability of farming systems in semi-arid Kenya, recording physical aspects of the environment together with land-management practices in four villages of Makueni District (Mbuvi, 2000). These span a range of settlement times from the 1950s to the 1970s. A major problem for farming in this region has been the frequency with which it has been affected by drought. Droughts reduce the returns that farmers receive and, consequently, provide little incentive for investment in soil fertility.

Physical Attributes

The climate of this region is characterized by two rainy periods: the short rains, which deliver most precipitation in October - December; and the long rains of March - May. However, superimposed on the average annual rainfall (600 - 670 mm) is the periodic occurrence of drought (Gichuki, 2000). Annual mean temperature is in the range of 21 - 24 °C. Elevation is 800 - 1 600 m, and the natural vegetation is grassland and dense shrub-land or woodland. Fires have affected the area in the past and the grassland is used for grazing. The soils are mostly Ferralsols (Rhodic and Xanthic) and are naturally low in phosphorus (Mbuvi, 2000).

Farming Systems

Annual or multiple cropping is practised with occasional incorporation of one-year fallows, although the latter is becoming less common. Each year covers two cropping seasons. The main crops are maize and pulses, with millet and sorghum recommended as drought crops. Yields vary considerably between years depending on rainfall; on average maize will yield 1 tonne/ha, although some modern varieties yield 4 tonnes/ha of grain (Mbogoh, 2000).

The animal population is not large because it is difficult to supply adequate feed during periods of drought. Consequently, manure is in short supply and highly valued. Scarcity means that its application is often rotated. Little investment is made in the grazing lands, and the animals are usually kept in pens during the wet season. Very little fertilizer is used, especially as drought causes "burning" of the crops. Crop residues are burnt, fed to animals or ploughed into the soil. Tillage is accomplished by using simple ridging ploughs pulled by oxen; hand hoeing and digging are also common. Labour is in short supply and a constraint for farmers. A range of crop-residue management and tillage techniques are used to conserve moisture and to protect against soil erosion (Pretty, Thompson and Kiara, 1995; Gichuki, 2000).

Woodland is being cleared, although selective clearance is often practised to save those species that provide useful products. Most farmers also plant trees, especially fruit trees and others such as mulberry for silk production (Gichuki, 2000). The study villages lie along a gradient of precipitation, decreasing from Kymausoi through Kaiani and Darjani to Athi Kamunyuni. CENTURY was run to equilibrium using a grassland - tree scenario with grass fires every ten years and major fires every 30 years.

Darjani

The primary crops for this settlement are millet, sisal, cowpea and sorghum and the natural pasture can support low intensity grazing. Settlement took place in the 1960s. The average soil carbon content of bush soils is 37.6 ± 7.5 tonnes/ha, and for cultivated soils it is 33.7 ± 2.8 tonnes/ha (Mbuvi, 2000).

Once run to equilibrium, CENTURY gave a value for soil C of 31.0 tonnes/ha in 1964. This is just within the range expected for bush soils. The model was then run for a further 35 years, reproducing the current land management. This commenced with a six-year cycle of alternate maize and millet crops for four years followed by fallow for two years. Cultivation was accomplished using a steel ridging plough and hand weeding was included. Crop residues were burnt. In 1971, FYM additions commenced, averaging 1.5 tonnes/ha over the six-year cycle as soil fertility and conservation measures began to be adopted. In 1983, the fallow period was reduced to one year out of six, and by 2000 the modelled soil C was 24 tonnes/ha. This reduction of 7 tonnes/ha from the level modelled for the uncultivated bush soil is greater than the decline of 5 tonnes/ha measured in the field. Continuation of this landmanagement practice is predicted to lead to a further loss of 1.3 tonnes C/ha by 2050.

RothC was run to equilibrium using the current level of bush soil C and then used to predict the effect of land management using plant inputs calculated by CENTURY. RothC predicted soil C to be 34.6 tonnes/ha in 1999, which is within 1 tonne/ha of the measured value. In the next 50 years, RothC predicts a further decline of 0.6 tonnes C/ha.

Effect of Organic Inputs

Adding an average 4.5 tonnes/ha/ year FYM over the 6-year croppingfallow cycle leads to a CS rate of 0.5 tonnes/ha/year. If the crop residues are returned to the soil rather than burnt, a further 4.9 tonnes C/ha can be accumulated by 2050, representing a sequestration rate of 0.6 tonnes/ha/year. RothC predicts similar increases. Increasing the FYM input to 6.75 tonnes/ha/year will increase the sequestration rate to 0.9 tonnes/ ha/year.

Effect of Inorganic Fertilizer

Replacing all organic inputs with inorganic fertilizer (100 kg N/ ha/ year, Scenario 4) results in only a moderate increase in soil CS (0.09 tonnes/ha/year). However, the quantity of N applied is equivalent to about five times that which is added in FYM for the current practice scenario. If the fallow period is removed and only inorganic fertilizer added (Scenario 3), the system behaves very much like the current practice scenario in spite of the additional N.

Summary

The importance of FYM as an organic input is demonstrated, as is the return of crop residues to the soil rather than burning them. Although addition of inorganic fertilizer can increase CS, the increase in C per unit of N added is much less efficient than if FYM is added. The indirect energy embodied in manufactured N fertilizer is an additional carbon cost.

Kaiani

The settlement of Kaiani has a farming system similar to that at Darajani. CENTURY underestimated the current bush content of soil C, yielding nearly 32 tonnes/ha compared with measured values of 41.5 ± 3.1 tonnes/ha.

CENTURY was then run with a scenario to reflect the last 40 years of cultivation using a millet - cowpea system with grazing of plant residues and 4.5 tonnes/ha manure applied once in the six-year cycle. There was one year of fallow.

Cultivation is calculated to have reduced soil C to 28.1 tonnes/ha by 2000, which compares with a measured value for cropped soil of 30.5 ± 4.8 tonnes/ha. Soil C is then predicted to decrease by slightly more than 1 tonne/ha in the next 50 years.

Table: *Total soil carbon for Kaiani settlement*

*Scenario**	*CENTURY*			*RothC*		
	2000	*2050*	*% change*	*2000*	*2050*	*% change*
	(tonnes/ha)			*(tonnes/ha)*		
1	28.07	26.81	-4.5	42.32	43.89	3.9
2		41.4	47.5		62.82	48.8
3		42.73	52.2		65.16	54.3

Source: CENTURY and RothC.

Table: *Scenarios for modelling land management practices, Kaiani settlement.*

Scenario	*Land management*
1	Current practice
2	FYM 2 tonnes/ha/year
3	FYM 2 tonnes/ha/year, plant residues 0.3 tonnes/ha/year

Table: *Total soil carbon for Kymausoi settlement*

*Scenario**	*CENTURY*			*RothC*		
	2000	*2050*	*% change*	*2000*	*2050*	*% change*
	(tonnes/ha)			*(tonnes/ha)*		
1	17.76	16.82	-5.3	35.69	34.79	-2.5
2		20.18	13.6		43.4	21.6
3		27.7	56.0		58.86	64.9

Source: CENTURY and RothC.

After parametization with the carbon content for bush soils, RothC shows very little effect of cultivation when using plant inputs calculated by CENTURY. Unlike CENTURY, this model predicts a slight rise in soil C by 2050.

Effect of Organic Inputs

Modest increases in the application rate of FYM (0.75 tonnes/ha/ year to 2 tonnes/ha/year) result in quite marked accumulation of soil C. Adding additional plant residues further increase the sequestration rate to 0.3 tonnes/ha/year. RothC predicts very similar proportional increases in soil C.

Summary

The models suggest that the current system is at or near steady state. It is possible to achieve reasonable rates of CS (0.3 tonnes/ha/ year) with modest increases in organic manure inputs.

Kymausoi

Kymausoi is situated in a marginal cotton zone where maize, pigeon pea and sisal are also grown. Cattle ranching is practised and the village was settled in the 1950s. The average soil carbon content for bush soils here is 38.4 ± 4.8 tonnes/ha, while for cultivated soils the average is 33.5 tonnes/ha with a range of 17.4 - 38.9 tonnes/ha. The CENTURY agro-ecosystem model was run to equilibrium and gave a bush soilcarbon level of 28.5 tonnes/ha in 1955, which is below the average concentration for soils in this area. The model was then run to reflect the farming system since settlement commenced. This includes continuous maize cropping with grazing of the crop residues, and an average manure application rate of 0.75 tonnes/ha/ year. Soil conservation and fertility management commenced in the late 1960s, the average annual manure application rate being increased to 1 tonne/ ha to reflect this. CENTURY estimated soil C to be 17.8 tonnes/ha by 2000, which is at the lower end of measured values. A decline of less than 1 tonne C/ha is predicted to occur in the next 50 years.

Table: *Scenarios for modelling land management practices, Kymausoi settlement.*

Scenario	Land management
1	Current practice
2	Plant residues 0.3 tonnes/ha/year
3	FYM 1.5 tonnes/ha/year, plant residues 0.6 tonnes/ha/year
4	FYM 1.5 tonnes/ha/year, plant residues 0.6 tonnes/ha/year, legume (cowpea)
5	FYM 2 tonnes/ha/year, plant residues 0.6 tonnes/ha/year, legume (cowpea)
6	FYM 4 tonnes/ha/year, plant residues 0.6 tonnes/ha/year, legume (cowpea)

Table: *Total soil carbon for Athi Kamunyuni settlement*

*Scenario**	*CENTURY*			*RothC*		
	2000	*2050*	*% change*	*2000*	*2050*	*% change*
	(tonnes/ha)			*(tonnes/ha)*		
1	23.75	23.45	-1.3	29.68	28.29	-4.7
2		29.09	22.5		31.40	5.8
3		29.77	25.3		42.76	44.1

Source: CENTURY and RothC.

After being run to equilibrium for current bush soils and then parametized with plant inputs from CENTURY, RothC calculated a reduction of 3.4 tonnes/ha in soil C resulting from cultivation. Like CENTURY, it suggests a further decline of less than 1 tonnes/ha by 2050.

Effect of Organic Inputs

Adding additional plant residues and increasing the FYM application rate to 1.5 tonnes/ha/year will increase soil carbon levels by at least 50 percent.

Effect of Legume Crop

Introducing a leguminous crop into the system, such as cowpea, can make a significant improvement in CS, increasing the rate from 0.2 to 0.3 tonnes/ha/year at the same rates of FYM application.

Summary

This example illustrates the advantage of including leguminous crops in the rotation. A farm system using 4 tonnes/ha/year of FYM, maintaining plant residues in the field, and with a legume in the rotation, can accumulate 0.7 tonnes/ ha/year. The current practice is reducing soil carbon stocks.

Athi Kamunyuni

This village is situated in a lowland agro-ecological zone. Ranching will only support a very low grazing density and the area is at the limit for rainfed production of millet, cowpea and sisal. The village was established in the 1970s. CENTURY was parametized for the commencement of cropping/grazing during the last 30 years. Millet was grown and an average of 0.75 tonnes/ha manure applied during the last six years.

The value for soil C modelled by CENTURY matched the current bush level (33.2 ± 3.2 tonnes C/ha). When settlement commenced, CENTURY calculated a decline in soil C of some 9 tonnes/ha over 25 years. Again, this is close to the actual measurements that show cultivated soils on average contain 24.0 ± 3.0 tonnes C/ha. Continuing this scenario into the twenty-first century suggests that the system has almost reached a new steady state. RothC calculates a smaller effect of cultivation on soil C, predicting a value of 29.7 tonnes C/ ha in 2000, and a slight decline over the next 50 years.

Table: *Scenarios for modelling land management practices, Athi Kamunyuni settlement.*

Scenario	*Land management*
1	Current practice
2	Grazing
3	FYM 1.25 tonnes/ha/year
4	FYM 2.25 tonnes/ha/year
5	FYM 2.27 tonnes/ha/year, fallow reduced to 1 year
6	FYM 3.3 tonnes/ha/year
7	FYM 3.9 tonnes/ha/year, plant residues 0.3 tonnes/ha

Effect of Organic Inputs

Increasing the current average application rate of FYM from 0.75 tonnes/ha/year to 1.25 or 2.25 tonnes/ha/year (Scenarios 3 and 4) increases rates of CS by 0.12 - 0.37 tonnes/ha/year. Further additions of organic inputs, if available, could yield an increase in CS rates of up to 0.6 tonnes/ha/year.

Effect of Fallows

Reducing the fallow period from two years to one year (Scenario 5) has very little effect on soil C providing the organic inputs are maintained. Using this location solely for ranching would result in an increase in soil C as the system returned to similar conditions that existed before cultivation commenced (Scenario 2). The rate of CS is very similar to Scenario 3, where an average of 1.25 tonnes/ha FYM was added annually. RothC predicts a much smaller increase in soil C on return to ranching. This reflects the fact that this model initially calculated a much smaller decline in total soil C following the commencement of cultivation.

Summary

The level of soil C in this recently settled system could be restored to pre-settlement levels and then raised higher by the addition of FYM. Conditions at this location are not ideal for cropping and a return to a grazing-only system should also restore soil C to its pre-settlement level.

Conclusions from Kenya Cases

The modelling of farm data from four communities in the semi-arid Makueni District again shows that carbon stocks can increase when a variety of technologies and practices already available to farmers are used. Modest inputs of organic material in the form of FYM and plant residues can lead to CS, particularly where systems are currently at or near steady state for soil carbon stocks.

Removal of fallow periods from existing systems results in losses of 0.1 tonnes C/ha/year. Inorganic fertilizers are again an inefficient choice for plant nutrients when soil C is a concern. Burning plant residues is not desirable. The combination of legumes in rotations, 2 - 4 tonnes/ha/year of FYM in addition to 0.6 tonnes/ha/year of plant residues, results in the highest rate of CS in all the dryland cases - 0.7 tonnes C/ha/year. At lower levels of FYM but with maintenance of fallows, combined with legumes in rotation, increases in soil C of 0.3 - to 0.4 tonnes/ha/year can be achieved.

Case Study 4 - Argentina - Tucuman, Catamarca and Cordoba Provinces

In recent years, Argentina has experienced a rapid growth in the adoption of reduced and no-tillage systems, especially in dryland regions. This change has been brought about by a deterioration in soil quality and associated crop yields. Many local soils are not suited to the heavy tillage and cropping practices introduced by European settlers.

The Argentine Pampa now has very little natural vegetation. Xerophitic vegetation such as *Prosopis algarrobilla* and *Larrea divaricata* can still be found in the most arid areas. Agricultural practices commenced with the arrival of colonists in the sixteenth century. Ungulates were introduced to graze the grasslands, which have now been mostly re-sown. Very few trees remain except around farmsteads. Wheat was initially cultivated and row-crop production has increased with time. In many parts, grazed pasture was dominant until the 1990s but since then there has been a marked increase in the cultivation of summer annuals, such as maize, sunflower and soybean (Diaz-Zorita, Duarte and Grove, 2002). The Argentine Pampa has been recognized as a region with potential for increased production, if soils can be improved (Alvarez, 2001).

Crop yields have declined in many areas. These declines have been correlated closely with a reduction in SOM content (Diaz-Zorita, Duarte and Grove, 2002). This has prompted the need for change in existing land-management practices. The negative effects of heavy tillage on SOM led to the commencement of no-tillage experiments in the 1960s, in an attempt to produce a more sustainable agricultural system. Now some 13 million ha, or about half of the agricultural area in Argentina, is under some form of reduced-tillage system. Fertilization of crops is primarily achieved through the use of inorganic fertilizers, with organic material tending to be conserved for use in horticultural farming systems.

No-tillage Modelling Studies

Three case studies are reviewed to model soil C under a variety of conventional and no-tillage systems in Tucuman, Catamarca and Cordoba Provinces.

Monte Redondo, Tucuman Province

This is a semi-arid area that naturally supports xerophitic vegetation. The agricultural practices include grazed prairie lands and row cropping, the two systems often being rotated. The studied site consists of cropping for seven years followed by four years of prairie grassland. The crop sequence is wheat/soybean, maize, soybean, wheat/soybean, maize, soybean, wheat, and four years of prairie. Both conventional tillage and no-tillage cultivation are practised. In the tillage system, disc and chisel plough are used for soil preparation, while the no-tillage system uses the same cropping sequence without tillage.

After an equilibrium phase of grassland and trees, with fire every 60 years, CENTURY was parametized to run with improved prairie grassland from the midnineteenth century. The model predicts that the prairie system is losing C at the rate of 0.06 tonnes/ha/year under the current regime, but at stocks of 55.4 tonnes C/ha, it is overestimating the current measured level of 48.8 tonnes C/ha.

Cultivation is scheduled to commence in the 1950s. Fertilizer applications of 110 kg/ha urea begin in 1980. Two scenarios compare the effect of conventional tillage with that of no-tillage. The model predicts a consistent difference between the two systems of less than 3 tonnes C/ha. This is less than the difference of 6 tonnes C/ha currently measured in the field. CENTURY predicts that this difference between the two systems will be maintained into the future, although both systems continue to lose soil C. The pattern of fall and rise in the soil C curve occurs because the crop part of the rotation results in soil C loss, while the return to prairie increases soil C.

Table: *Total soil carbon for Monte Redondo*

*Scenario**	*CENTURY*			*RothC*		
	2000	*2050*	*% change*	*2000*	*2050*	*% change*
	(tonnes/ha)			*(tonnes/ha)*		
Prairie	55.40	54.50	-1.6	43.95	43.21	-1.7
Tillage	40.43	39.27	-2.9	32.57	27.58	-15.3
No-tillage	43.64	42.23	-3.2	35.33	33.58	-5.0
5		48.23	4.4		41.89	18.6

Source: CENTURY and RothC.

Table: *Scenarios for modelling land management practices, Monte Redondo.*

Scenario	*Land management*
1	Conventional tillage, inorganic fertilizer
2	No-till, inorganic fertilizer
3	No-till, FYM 1.5 tonnes/ha/year, inorganic fertilizer
4	No-till, green manure 10 tonnes/ha/crop, inorganic fertilizer
5	No-till, FYM 1.5 tonnes/ha/year, green manure 10 tonnes/ha/crop, no inorganic fertilizer
6	No-till, FYM 3.3 tonnes/ha/crop, no inorganic fertilizer

RothC calculates lower levels of soil C but predicts a greater differential of 6 tonnes/ha by 2050, mainly through a higher loss of C from the tilled system.

Effect of No-tillage and Organic Additions

Additions of green manure (10 tonnes/ha/crop) or FYM (1.5 tonnes/ha each cropping year) to the no-tillage system both lead to increases in CS. A combination of these inputs without inorganic fertilizer yields a similar result. Cessation of inorganic fertilizer usage and using FYM as a replacement source of N results in the highest rate of CS (Scenario 6), 0.1 tonnes/ha/year.

Summary

Both models register the improvement that no-tillage has on soil carbon content. However, if the decline in soil carbon content is to be reversed, additional inputs of organic matter are required - either from

FYM or use of green manures in the rotation. An increase in prairie in the rotation will also increase soil carbon stocks.

Santa María River Valley, Catamarca Province

This is an arid region, with mean annual rainfall of 400 mm and temperatures in the ranging of 7 - 32 °C. The native vegetation is xerophitic consisting of creosote bush scrub (*Larrea divaricata*) and trees such as *Prosopis algarrobilla*. In cultivated areas, vines are grown and crops include alternate plantings of red pepper and barley. Measured soil carbon stocks are high in this district at 3.9 percent.

Table: *Total soil carbon for Santa Maria*

*Scenario**	*CENTURY*			*RothC*		
	2000	*2050*	*% change*	*2000*	*2050*	*% change*
	(tonnes/ha)			*(tonnes/ha)*		
Prairie	37.11	35.75	-3.7	70.68	74.71	5.7
Tillage	21.39	18.11	-15.3	51.46	47.36	-8.0
No-tillage		24.00	12.2		53.53	4.0
3		30.49	42.5		58.71	14.1

Source: CENTURY and RothC.

After reaching equilibrium with natural vegetation, CENTURY was run with a prairie scenario from the mid-nineteenth century and cultivation commenced in the late 1950s using conventional tillage (disc plough). A cotton - barley rotation commenced in 1980. CENTURY calculates that soil C had decreased to two-thirds of its initial value by 2000. Measurements in the region confirm declines in soil C in cultivated areas of 33 - 66 percent. Therefore, the model estimate is at the top of this range. However, the system is predicted to reach a new steady state and the decline in soil C is estimated to be 3 tonnes/ha over the next 50 years.

RothC, parametized with the higher, measured level of soil C and using quantities of plant inputs calculated from CENTURY shows a proportionately smaller effect of the current cultivation practice on soil C.

Effect of no-tillage

Adopting a no-tillage system not only halts the loss of soil C predicted by CENTURY. It also leads to a low sequestration rate of 0.05 tonnes/ha/ year in the next 50 years. The RothC model is less sensitive to this scenario.

Effect of Organic Additions

Additions of FYM and green manure both lead to marked increases in CS rates of 0.18 - 0.25 tonnes/ha/year. RothC again shows a smaller effect.

Effect of Inorganic Fertilizer

Replacing inorganic fertilizer with FYM promotes CS (0.22 tonnes/ha/year). Combining the green manure and FYM applications gives the best carbon accrual rate, 0.29 tonnes/ha/year. Inorganic N fertilizer is inefficient owing to the high energy cost of manufacture.

Table: *Scenarios for modelling land management practices, Santa Maria.*

Scenario	*Land management*
1	Conventional tillage, inorganic fertilizer
2	No-till, inorganic fertilizer
3	No-till, FYM 1.5 tonnes/ha/year, inorganic fertilizer
4	No-till, FYM 3.3 tonnes/ha/crop, no inorganic fertilizer
5	No-till, green manure 10 tonnes/ha/crop, inorganic fertilizer
6	No-till, FYM 1.5 tonnes/ha/year, green manure 10 tonnes/ha/crop, no inorganic fertilizer

Summary

Both models suggest that adopting no-tillage will halt the decline in soil C. However, to increase CS, higher organic additions are necessary (green manures and FYM), which can be used to replace the inorganic fertilizer applications.

Cordoba Province, Buenos Aires Province, and La Pampa Province

Following an equilibrium period with natural vegetation and subsequent prairie conditions from the mid-1800s, CENTURY was parametized with a cultivation regime commencing in the 1950s. This included a fouryear cropping (wheat - soybean - maize - soybean) and a four-year prairie cycle with inorganic fertilizer applications (100 kg/ha urea) starting in 1985. In 1987, a rotated and a nonrotated cropping system was applied that is similar to cultivation practices occurring in the field. The rotated crop sequence was winter forage - soybean - maize - soybean - wheat - soybean - maize - four years of prairie and winter forage - wheat - soybean - maize - soybean - maize - prairie. The non-rotated crop sequence was similar but without the prairie interludes.

Modelled results for the nonrotated crop system show an initial steep fall in soil C to 37 tonnes/ha in 2000 and further losses over subsequent years. RothC estimates a proportionately larger fall from a higher base. The rotated crop - prairie system does not show the same sharp decline in soil C as the nonrotated system did and oscillates around a level of slightly more than 40 tonnes/ha. However, RothC estimates a slightly larger proportionate decline in C for the rotated crop system. Both models calculate smaller differences between the rotated and non-rotated cropping systems in 2000 compared with the difference of 8.5 tonnes/ha measured in the field.

Table: *Total soil carbon for rotated and non-rotated plots, modelled with CENTURY and RothC.*

*Scenario**	*CENTURY*			*RothC*		
	2000	*2050*	*% change*	*2000*	*2050*	*% change*
	(tonnes/ha)			*(tonnes/ha)*		
Non-rotated plots	37.22	32.74	-12.0	50.61	41.17	-18.7
Rotated plots	42.47	40.91	-3.7	54.62	49.53	-9.3
4		45.54	7.2		62.86	15.1

Table: *Scenarios for modelling land management practices, Santa Maria.*

Scenario	***Land management***
1	Non-rotated plots, inorganic fertilizer
2	Rotated plots, inorganic fertilizer
3	Rotated plots, no-till, inorganic fertilizer
4	Rotated plots, no-till, FYM 1.5 tonnes/ha/year, green manure 10 tonnes/ha/crop, no inorganic fertilizer
5	No-till, FYM 1.5 tonnes/ha/year, inorganic fertilizer
6	No-till, green manure 10 tonnes/ha/crop, inorganic fertilizer
7	No-till, FYM 3.3 tonnes/ha/crop, no inorganic fertilizer

Effect of no-tillage

The adoption of a no-tillage regime for the rotated-plots system increases soil C by 2.5 tonnes/ha in the next 50 years, representing a CS rate of 0.02 tonnes/ha/year. The rate increases to about 0.1 tonnes C/ ha/year if green manures and FYM are used instead of fertilizers.

Effect of Organic Inputs

Additions of green manure and FYM with or without inorganic fertilizer can lead to CS rates of 0.06 - 0.13 tonnes/ha. Organic material can replace inorganic fertilizer successfully.

Summary

The inclusion of prairie interludes in the cropping system is an important factor for reducing the decline in soil C. However, the models show that no-tillage and organic inputs are required if C is to be sequestered in this system.

Conclusions from Argentina Cases

The modelling of farm data from three dryland provinces of Argentina shows that carbon stocks have fallen substantially since the prairies were opened up for cultivation. At all three locations, there have been sharp falls in soil carbon stocks, with losses of about 15 tonnes/ha.

However, the adoption of no-tillage systems in recent years has halted these declines and, on their own, resulted in small annual increases in soil C of the order of 0.02 tonnes/ ha/year. Rotations with significant periods for return to prairie grassland (e.g. 4 years in 11) result in further increases in soil C. The highest rates of sequestration (0.1 - 0.25 tonnes/ha/year) occur when no-tillage systems also include cultivation of green manures and additions of FYM.

Case Study 5 - Senegal - Old Peanut Basin

The Senegal study area, the "Old Peanut Basin", is located in the west-central part of the country. The climate is semi-arid with annual precipitation ranging of 350 - 700 mm. Almost all arable land is used for rainfed agriculture, with millet, groundnuts, sorghum and cowpea as major crops. The rainy season usually lasts from July to September/ October. However, both spatial and temporal variation in rainfall are high and episodic crop failures are not uncommon. The natural vegetation, including *Faidherbia albida* and various other tree species, has become heavily degraded, primarily because of a long agricultural history and increasing population pressure.

As described in Tschakert (2004b), CENTURY model simulations suggest that soil C in the study area decreased from 20.1 tonnes C/ha under a native savannah environment in 1851 to 11.9 tonnes C/ha in 2001. This indicates an annual loss of soil C of 0.055 tonnes C/ha/year. Tree C declined from 33.6 tonnes/ha to 4.2 tonnes/ha, corresponding to an annual decrease of 0.2 tonnes C/ha/year.

Under improved management conditions (assuming a 50 - year period), soil C could increase by 0.3 - 13.5 tonnes/ha, or 0.006 - 0.27 tonnes/ha/year. C gains in trees could be tripled (from 4.2 to 11.8 tonnes/ ha), assuming a conversion of croplands to grassland - tree plantations. Given the fact that most of the C gains are achieved in the first 25 years, annual increases for this time period range from 0.02 to 0.43 tonnes C/ha/year, which is higher than the estimates provided by Lal, Hassan and Dumanski (1999). Under poor management, in this case an annual millet - sorghum rotation with no inputs and permanent browsing and pruning of tree resources, both soil and tree C continue to drop, reaching an absolute minimum level of 7.9 tonnes/ha and 0.6 tonnes/ ha, respectively.

The various improved management practices with the anticipated changes in soil C over two time periods (2002 - 2026 and 2027 - 2050) as discussed by Tschakert (2004b). As illustrated, considerably higher gains in soil C can be achieved in the first 25 years, except for the tree-plantation scenario. However, in some cases, these gains cannot be sustained in the second twenty 25-year period, and losses would have to be expected if no additional inputs occurred. This is particularly true for the fallow scenarios.

***Table:** Effects of land management practices or land use on carbon sequestration potential in the Old Peanut Basin, Senegal*

Technological options*	***Change in soil C (tonnes/ha/year) 2001 - 2026 (first 25 years)***	***Change in soil C (tonnes/ha/year) 2027 - 2050 (second 25 years)***
Compost (2t)	0.02	-0.01
Conversion of croplands to grasslands + grazing	0.06	0.02
3-year fallow + 2 tonnes manure in rotation with 4 years cropping	0.14	-0.05
Cattle manure (4 tonnes)	0.10	0.01
Conversion of croplands to grasslands	0.17	0.04
Cattle manure (4 tonnes) + min. fertilizer (250kg on millet; 150kg on	0.12	0.01
groundnuts)		
Sheep manure (5 tonnes)	0.13	0.01
3-year fallow + Leucaena prunings (2 tonnes) in rotation with 4 years	0.18	-0.05
cropping		
Conversion of croplands to grasslands with tree protection	0.10	0.04
Sheep manure (10 tonnes)	0.17	0.01
10-year fallow + 2 tonnes manure in rotation with 6 years cropping	0.25	-0.04
10-year fallow + Leucaena prunings (2 tonnes) in rotation with 6 years cropping	0.25	-0.04
Conversion of croplands to grasslands + tree plantation (*Faidherbia albida*)	0.23	0.21
Agricultural intensification (improved millet, manure, Leucaena prunings, min. fertilizer, animal traction, 1-year fallow)	0.43	0.11

Agricultural intensification (improved millet, manure, Leucaena prunings, min. fertilizer, animal traction, 1-year fallow) 0.43 0.11

*All technological options calculated for 1 ha. Annual rotation between millet and groundnuts for cropping scenarios.

Source: Tschakert (2004b).

Chapter 8

Management Steps to Improve Soil Quality

A basic soil audit is the first and sometimes the only monitoring tool used to assess changes in the soil. Unfortunately, the standard soil test done to determine nutrient levels (P, K, Ca, Mg, etc.) provides no information on soil biology and physical properties. Yet, most of the farmer-recognized criteria listed on the first page of this publication for healthy soils include, or are created by, soil organisms and soil physical properties. A better appreciation of these biological and physical soil properties, and how they affect soil management and productivity, has resulted in the adoption of several new soil health assessment techniques.

The USDA Soil Quality Institute provides a Soil Quality Test Kit Guide developed by John Doran and associates at the Agricultural Research Service's office in Lincoln, Nebraska. The kit was designed for field use. Components necessary to build a kit include many items commonly available such as pop bottles, flat bladed knives, a garden trowel, and plastic wrap. Also necessary to do the tests is some equipment usually not locally available such as hypodermic needles, latex tubing, a soil thermometer, an electrical conductivity meter, filter paper, and an EC calibration standard. The kit allows the measurement of water infiltration, water holding capacity, bulk density, pH, soil nitrate, salt concentration, aggregate stability, earthworm numbers, and respiration.

A greatly simplified and quick soil quality assessment is also available at the Soil Quality Institute's web page by clicking on "Getting to Know your Soil," near the bottom of the homepage. This greatly simplified method involves digging a hole and making some observations. Here are a few of the procedures shown at this website: Dig a hole 4 to 6 inches below the last tillage depth and observe how hard the digging was.

Inspect plant roots for lots of branching and fine root hairs or a balled up condition. A lack of fine root hairs indicates oxygen deprivation, while sideways growth indicates a hardpan. The process goes on to tell about earthworms, smelling the soil, and assessing the aggregation.

A cropland-monitoring guide has been published by the Centre for Holistic Management. The monitoring guide contains a set of soil health indicators that are measurable in the field. No fancy equipment is needed to make the assessments described in this monitoring guide. In fact, all the equipment is cheap and locally available on almost any farm. Simple measurements can help determine the health of croplands in terms of the effectiveness of the nutrient cycle, water cycle, and the diversity of some soil organisms. Some of the assessments you can make using this guide are living organisms, aggregation, water infiltration, ground cover, and earthworms. The monitoring guide is easy to read and understand, and comes with a field sheet to record observations. It is available for $12 from the Centre for Holistic Management.

Some quick ways to identify a healthy soil include feeling it and smelling it. Grab a handful and take a whiff. Does it have an earthy smell? Is it a loose, crumbly soil with some earthworms present? Dr. Ray Weil, soil scientist at the University of Maryland describes how he would make a quick evaluation of a soil's health in just 5 minutes.

Look at the surface and see if it is crusted, which tells something about tillage practices used, organic matter, and structure. Pushing a soil probe down to 12 inches, lift out some soil and feel its texture. If a plow pan were present it would have been felt with the probe. Turn over a shovelful of soil to look for earthworms and smell for actinomycetes, which are microorganisms that help compost and stabilize decaying organic matter. Their activity leaves a fresh earthy smell in the soil.

Two more easy observations are to count the number of soil organisms in a square foot of surface crop residue and to pour a pint of water on the soil and record the time it takes to sink in. Comparisons can be made using these simple observations along with Ray Weil's evaluation above to determine how farm practices affect soil quality.

Simple Erosion Test

This test demonstrates the value of ground cover. Tape a white piece of paper near the end of a 3-foot-long stick. Hold the stick in one hand so as to have the paper end within 1 inch of a bare soil surface. Now pour a pint of water onto the bare soil within 2-3 inches of the white paper and observe the soil accumulation on the white paper. Tape another piece of white paper to the stick and repeat the operation, this time over soil with 100% ground cover, and observe the accumulation of soil on the paper. Compare the two pieces of paper. This simple test shows how effective ground cover can be at preventing soil particles from detaching from the soil surface.

Detachment of soil particles occurs when falling rainwater collides with bare ground. After enough water builds up on the soil surface, following detachment, overland water flow transports suspended soil down slope. Suspended soil in the runoff water abrades and detaches additional soil particles as the water travels overland. Preventing detachment is the most effective point of erosion control because it keeps the soil in place. Other erosion control practices which seek to slow soil particle transport and cause soil to be deposited before it reaches the stream are less effective at preventing erosion. These latter practices are the ones typically implemented (terraces and diversions). Terraces, diversions and many other erosion "control" practices are basically unnecessary if the ground stays covered year round. For erosion prevention, a high percentage of ground cover is a good "early warning" indicator of success, while bare ground indicates a high risk of erosion (Croplands monitoring guide, 98). Muddy runoff water and gullies are "too-late" indicators. The soil has already eroded by the time it shows up as muddy water and it's too late to save soil already suspended in the water.

Tools and Techniques to Build Soil

Can a cover crop be worked into your rotation? How about a high-residue crop or perennial sod? Are there economical sources of organic materials or manure in your area? Are there ways to reduce tillage and nitrogen fertilizer? Where feasible, bulky organic amendments may be added to supply both organic matter and plant nutrients. It is particularly useful to account for nutrients where organic fertilizers and amendments are utilized. Start with a soil test and a nutrient analysis of the material you are applying. Knowing the amount of nutrients needed to supply the crop to be grown guides the amount of amendment applied and can lead to significant reductions in fertilizer purchase. The nutrient composition of organic materials can be variable, which is all the more reason to determine the amount you have with appropriate testing. In addition to containing the major plant nutrients, organic fertilizers can supply many essential micronutrients. Proper calibration of the spreading equipment is also important to ensure accurate application rates.

Animal Manure

Manure is an excellent soil amendment, providing both organic matter and nutrients. Typical rates for dairy manure would be 10 to 30 tons per acre or 4,000 to 11,000 gallons of liquid for corn. At these rates the crop would get between 50 and 150 pounds of available nitrogen per acre. Additionally, lots of carbon would be added to the soil, resulting in no loss of soil organic matter. High crop residues grown from this manure application would also contribute organic matter.

However, a common problem with using manure as a crop nutrient source is that application rates are usually based on the nitrogen needs of the crop. Because some manures often have about as much phosphorus as they do nitrogen, this often leads to buildup of soil phosphorus. A classic example is chicken litter applied to crops that require high nitrogen levels, such as pasture grasses and corn. Broiler litter, for example, contains approximately 50 pounds of nitrogen and phosphorus and about 40 pounds of potassium per ton. A common fertilizer application for established fescue pasture would be about 50 pounds of nitrogen and 30 – 40 pounds of phosphorus per acre. If a ton of poultry litter were applied to supply the nitrogen needs of the fescue, an over-application of phosphorus would result. Several years of litter application can build soil phosphorus up to excessive levels. One easy answer is to this dilemma is to adjust the manure rate to meet the phosphorus needs of the crop and to supply the additional nitrogen with fertilizer or a legume cover crop.

Compost

Composting farm manure and other organic materials is an excellent way to stabilize their nutrient content. A significant portion of raw-manure nutrients are in unstable, soluble forms. Such unstable forms are more likely to run off if surface applied, or to leach if tilled into the soil. Therefore compost is not a good source of readily available plant nutrients like manures are. Compost releases its nutrients slowly, thereby minimizing losses. Quality compost contains more humus than its raw components because primary decomposition has occurred during the composting process. It also does not contribute the sticky gums and waxes that aggregate soil particles together as much as does raw manure because these substances are also released during the primary decomposition phase. Unlike manure, compost can be used at almost any rate without burning plants. In fact some greenhouse potting mixes contain 20 to 30% compost. Compost (like manure) should be analyzed by a laboratory to determine the nutrient value of a particular batch and insure its wise use. Composting also reduces the bulk of raw organic materials–especially manures which often have a high moisture content. However, while less bulky and easier to handle, composts can be expensive to buy. On-farm composting cuts costs dramatically compared with buying compost.

Cover Crops and Green Manures

Many types of plants can be grown as cover crops. Some of the more common ones include: rye, buckwheat, hairy vetch, crimson clover, subterranean clover, red clover, sweet clover, cowpeas, millet, and forage sorghums. Each of these plants has advantages over the others and their area of adaptability. Cover crops can maintain or increase soil organic matter if they are allowed to grow long enough to produce high herbage. All too often, people get in a hurry and take out a good cover crop just a week or two before it has reached its full potential. Hairy vetch or crimson clover can yield up to 2.5 tons per acre if allowed to go to 25% bloom stage. A mixture of rye and hairy vetch can produce even more.

In addition to the organic matter benefits, legume cover crops provide considerable nitrogen for crops that follow them. Consequently, the nitrogen rate can be reduced following a productive legume cover crop taken out at the correct time. For example, corn grown following 2 tons of hairy vetch should produce high yields of grain with only 50% of the normal nitrogen rate.

When small grains such as rye are used as cover crops and allowed to reach the flowering stage, additional nitrogen may be required to help offset the nitrogen tie-up caused from the high carbon addition of the rye residue. The same would be true of any high carbon amendment such as sawdust or wheat straw. Cover crops also suppress weeds, help break pest cycles, and through their pollen and nectar provide food sources for beneficial insects and honeybees. They can also cycle other soil nutrients making them available to subsequent crops as the green manure decomposes.

Humates

Humates and humic acid derivatives are a diverse family of products, generally obtained from various forms of oxidized coal. Coal-derived humus is essentially the same as humus extracts from soil but there has been a reluctance in some circles to accept it as a worthwhile soil additive. In part, this stems from a belief that only humus derived from recently decayed organic matter is beneficial. It is also true that the production and recycling of organic matter in the soil cannot be replaced by coal-derived humus. However, while sugars, gums, waxes and similar materials derived from fresh organic-matter decay play a vital role in both soil microbiology and structure, they are not humus. Only a small portion of the organic matter added to the soil will ever be converted to humus. Most will return to the atmosphere as carbon dioxide as it decays.

Many studies have shown positive effects of humates, while other studies have shown no such effects. Generally, the consensus is that they work well in low organic matter soils. In low amounts they do not produce positive results on soils high in organic matter. At high rates they may tie up soil nutrients.

There are many humus products on the market. They are not all the same. Humate products, should be evaluated in a small test plot for cost effectiveness before using. Sales people sometimes make exaggerated claims for their products.

Reduce Tillage

While tillage has become common to many production systems, its effects on the soil can be counter-productive. Tillage smoothes the soil surface and reduces natural soil aggregation and earthworm channels. Porosity and water infiltration are decreased following most tillage operations. Plow pans may develop in many situations. Tilled soils have much higher erosion rates than soils left covered with crop residue.

Due to all the problems associated with conventional tillage operations, acreage under reduced tillage systems is increasing on the American landscape. Any tillage system that leaves in excess of 30% surface residue is considered a “conservation tillage” system by USDA. Conservation tillage includes no-till, zero till, ridge-till, zone till, and some variations of chisel plowing and disking. These conservation till strategies and techniques allow for establishing crops into the previous crop’s residues, which are purposely left on the soil surface. The principal benefits of conservation tillage are the reduction of soil erosion and improved water retention in the soil, resulting in more drought resistance.

Additional benefits, which many conservation tillage systems provide, include reduced fuel consumption, flexibility in planting and harvesting, reduced labour requirements, and improved soil tilth. Two of the most common conservation tillage systems are ridge tillage and no-till. Ridge tillage is a form of conservation tillage that uses specialized planters and cultivators to maintain permanent ridges on which row crops are grown. After harvest, crop residue is left until planting time. To plant the next crop, the planter places the seed in the top of the ridge after pushing residue out of the way and slicing off the surface of the ridge top. Ridges are re-formed during the last cultivation of the crop. Often, a band of herbicide is applied to the ridge top during planting. With banded herbicide applications, two cultivations are generally used: one to loosen the soil and another to create the ridge later in the season. No cultivation may be necessary if the herbicide is applied broadcast rather than banded. Because ridge tillage relies on cultivation to control weeds and reform ridges, this system allows farmers to further reduce their dependence on herbicides, compared with either conventional till or strict no-till systems.

Maintenance of the ridges is key to successful ridge tillage systems. The equipment must accurately reshape the ridge, clean away crop residue, plant in the ridge centre, and leave a viable seedbed. Not only does the ridge-tillage cultivator remove weeds, it also builds up the ridge. Harvesting in ridged fields may require tall, narrow dual wheels to be fitted to the combine. This modification permits the combine to straddle several rows, leaving the ridges undisturbed. Similarly, grain trucks and wagons cannot be driven randomly through the field. Maintenance of the ridge becomes a consideration for each process.

Conventional no-till methods have been criticized for a heavy reliance on chemical herbicides for weed control. Additionally, no-till farming requires careful management and expensive machinery for some applications. In many cases, the spring temperature of untilled soil is lower than that of tilled soil. This lower temperature can slow germination of early-planted corn or delay planting dates.

Also, increased insect and rodent pest problems have been reported. On the positive side, no-till methods offer excellent soil erosion prevention and decreased trips across the field. On well-drained soils that warm adequately in the spring, no-till has provided the same or better yields as conventional till. A recent equipment introduction into the no-till arena is the so-called "no-till cultivator." These cultivators permit cultivation in heavy residue and provide a non-chemical option to post-emergent herbicide applications. Farmers have the option to band herbicide in the row and use the no-till cultivator to clean the middles as a way to reduce herbicide use.

Minimize Synthetic Nitrogen Fertilizer Use

If at all possible, add carbon with nitrogen sources. Animal manure is a good way to add both carbon and nitrogen. When nitrogen fertilizer is used, try to do it at a time when a heavy crop residue is going onto the soil, too. For example, a rotation of corn, beans, and wheat would do well with nitrogen added after the corn residue was rolled down or lightly tilled in.

Spring planted soybeans would require no nitrogen. A small amount of nitrogen could be applied in the fall for the wheat. Following the wheat crop, a legume winter-annual cover crop could be planted. In the spring, when the cover crop is taken out, nitrogen rates for the corn would be reduced to account for the nitrogen in the legume.

The addition of legume residue would also be adding carbon. Avoid continual hay crops accompanied by high nitrogen fertilization. The continual removal of hay accompanied by high nitrogen speeds the decomposition of soil organic matter. Heavy fertilization of silage crops, where all the crop residue is removed (especially when accompanied by tillage), speeds soil decline and organic matter depletion.

Continue to Monitor for Indicators of Success or Failure

As you experiment with new practices and amendments, continue to monitor the soil for changes using some of the tools discussed in the Assessing Soil Health and Biological Activity section. Several of these monitoring guides have data sheets you can use in the field to record data and use for future comparison after changes are made to the farming practices. Review the principles of sustainable soil management and find ways to apply them in your operation. If the thought of pulling everything together seems overwhelming, start with only one or two new practices and build on them.

Crop Production Management

Pre-broadcasting Treatment for Better Groundnut Harvest

Groundnut is usually planted by broadcasting. Before broadcasting the seeds they are treated in three steps. First the groundnut is dried in the sun and graded. Drying seeds before peeling off can improve seed vigour. Usually the skin of the peanut is peeled off 4 days before broadcasting and then large size seeds are selected for broadcasting.

Then the seeds are soaked in cold water for 12-14 hours followed by again soaking in lukewarm water (35° C) for 2 minutes. The soaked seeds are then put in a bucket and a wet grass film is used to wrap them closely. The bud will sprout after 20 hours.

The third step is dressing the seed by pesticide. This is done by evenly splashing ammonium molybdate desolved in lukewarm water on the seeds and then dressing the seeds with duojunling (a local pesticide) in order to prevent disease and pests. For 50 kg of seeds 100 g ammonium molybdate and 250 g of Duojunling with a 25% concentration is required.

Controlling Number of Tillers

Rice yield can be increased by 30-40% by controlling the number of tillers. In this method the water is drained from the paddy fields so that the soil is exposed to the sun and allowed to dry till fissures appear. The fields are then irrigated again and this procedure is repeated several times till the number of tillers is only 80% of normal tillers of a variety. If there is more rain in tropical areas and there are less sunny days, there is no need to irrigate again after draining water until two leaf stage. The paddy field should be kept dry and wet alternately. The method produces 20% more grain bearing ears compared to the rate without the use of this method.

Another method, which can increase rice yield by 30%, is by frequent irrigation to the depth of two leaves. After the two leaf stage the field should is kept wet and dry alternately. In fertile clay soils more sunshine is needed other wise in sandy and less fertile soils long exposure to sunshine is not needed.

Method of Preventing Cotton Plant with non-blooming Buds

A successful method to solve the problem of cotton plants bearing non-blooming buds is to spray on the leaves a 0.1 % boric acid and 0.2 % borax hot mix diluted in cold water. The rate of spray is 6,975 liters/ha. The spraying is done during seedling stage, bud stage and blooming stages, respectively.

Rice Blast Diagnosis

There are two methods for diagnosis of rice blast in advance. One method is to take 6 rice leaves with some spots on them from the field.

Put them in a bowl with a piece of wet cotton. A bowl is closed with a plastic cover. Put the bowl in an oven at 24°C for 24 hours. Then take the leaves out of the bowl and check whether there is a gray mould on the back of the leaf. If there is, then it can be confirmed that the rice plant in the field is suffering from blast.

Another method is to take the 6 rice leaves with some spots from the field. Bind them and hang them above hot water at 40°C in a bottle. Then close the bottle for 24 hours. After that, check at the back of the leaves. The appearance of gray mould on it shows that the rice field will be attacked by blast. Preventive measures should be taken as early as possible.

Inter-cropping Corn and Chili for Increased Production

Inter-cropping corn and chili can enhance land productivity. The corn production not only remains the same but the chili crop also gets good profit.

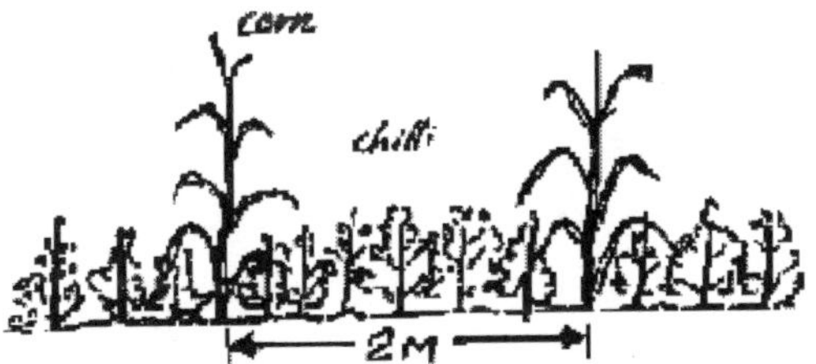

For land preparation, it is tilled to 25 cm depth and spread with organic manure. Now a days phosphorus and nitrogen (1:2/3) are also applied to the soil. The corn is planted at a row spacing of 100 cm × 15 cm with a plant density of 4500 rows per 0.15 ha. Two columns of chili are planted between corn in 30×30 cm spacing. The density of chili is at a rate of 4500 rows per 0.15 ha.

Now a day fertilizer is sprayed during crop growth. Weeding and pest control before corn cobs are formed require special attention. To promote chili to ripe old corn leaves should be removed during the growing stage of the corn. The heights of corn and chili plants are quite different. Lower leaves of corn shadow the chili, which is beneficial to it. On an average about 77 corn leaves are in 100-160 cm space, but chili occupies 60-80 cm space. This is a good arrangement to give sufficient light to chili. The corn production under this intercropping is same as of solo crop. Thus income from chili becomes extra. However, the method is suitable in summer season of semi-humid regions only as enough light and water are available at the same time.

Inter-cropping of Cotton and Melon (Cucumis Melo)

Compiled by Lui Xiaowen and Pong Shuping, Agriculture Extension Station, Zhibo, Sandong and Jia Fusheng and Sun Chuandian, Agriculture Extension Station, Bianhe, Sandong, China. For this inter-cropping arrangement, plant cotton in wide columns (80 cm) alternating with narrow columns (50-55 cm), with row spacing of 28-30 cm with a density is 3500 rows per 0.15 ha. Melon is planted in the space between wide columns of cotton with a row distance of 60-70 cm and a density of 800 rows per 0.15 ha for melon.

For good land preparation, in the winter season during plowing apply organic manure at a rate of 2500-3000 kg/ha, calcium phosphate 40-50 Kg/ha and urea 15-20 kg/ha. Then rake the soil for sowing in April in the tropical regions of Sandong province of China.

Pre-treat the cotton seeds in hot water at a temperature of 75°C for 2-3 days before sowing. The land is then covered with a plastic film after sowing. If the soil is dry it needs to be watered 1-2 liter/hole before sowing. Select melon seeds of medium variety with high yield. Melon is sown at the same time with cotton. Put 2-3 melon seeds in each hole and cover with 2 cm of soil.

Management

Seeding stage: By controlling the cotton density, it grows faster and promotes the growth of melon. It needs about 80 days for melon to ripen from the time of sowing. During this period cotton grows slowly and does not need too much water and fertilizer. This creates a good growing condition for melon.

Thinning out of the cotton seedling is required at two leaf stage. Control of the cotton pests such as aphid, thrips and cutworm etc also. The soil should be loosened to keep the moisture intact and to promote growth. If melon grows too fast, only 3-4 branches and 3-4 melons should be kept. Its head and other branches should be cut. It is necessary to control pests such as powdery mildew.

Middle stage: Melon ripens in June-July. It should be harvested on time, since cotton needs more nutrition at this time. Chemical liquid should be sprayed twice to foster the growth of strong cotton seedling after it has 5-8 buds. The first spraying should be done at the end of June and the second spraying should be in the middle of July. Take out plastic film in the last ten days of June, and work the soil to loosen it in the spaces between the cotton columns.

Cotton blooming and ball formation stage: It is necessary to spray 10-15 kg urea before the 10th of July and around the 20th-25th of July. The head of the plant should be cut at the same time when the pruning is done during the first week of August. Attention should be paid to drainage, to loosening of the soil and to timely pest control.

Since both cotton and melon are drought-resistant and have different growing periods, inter-cropping the two can have high economic benefits. Natural resources such as land, light and water are used very economically. This technology is well suited to the rainfed mountain areas. However the intercropping is not suitable for cotton and melon varieties that have long growing period.

Honey bee-a good guard against bird damage to grape: Birds like to eat ripe grapes. Honeybee is a good guard against this damage. If beehives are put in the grape orchards during their grape maturity period, it very successfully protects the grape from bird damage. Bees attack birds, when they come to eat grape. According to local experience about 30 beehives/ha of orchard is enough for complete damage control.

Soil and Water Conservation: Intermittent Slope Level Ditch

Shaanxi Province is located in the loess plateau mountain area. Soil erosion is very serious here. With a long history of erosion control practices, the local people have found that the intermittent slope level ditch is very useful for erosion control and land utilization. It has all the advantages of intermittent bench terraces, silting dam (locally called Yudiba), and fish-scale pits. It helps to intercept sediment, store water and at the same time increases fertility.

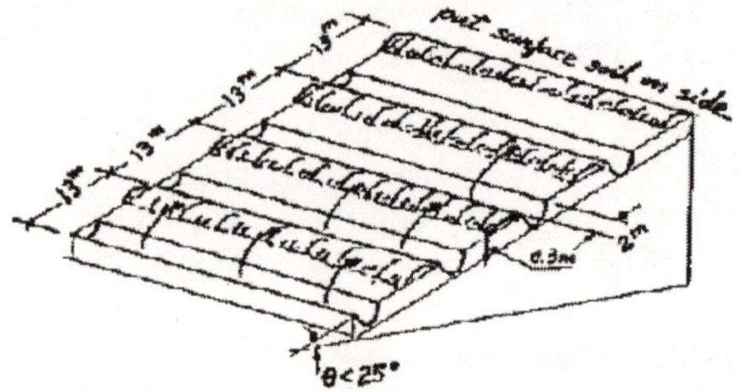

On slopes below 25°, the ditches are constructed 13 m apart along the contour,. The distance between the ditches is narrowed on slopes more than 25°. A ditch of 2 m width and 30 cm depth is first excavated and disposed on the inner side (upper slope); then on the outer side the earth is excavated to a depth of 1 m and the excavated material is disposed on the lower side for a ridge.

The ridge should be 0.7 m higher than the original slope surface with a bottom width of 1 m and top width of 0.5 m. The earth should then be rammed. Finally the excavated surface soil is refilled in to the ditch. The refilling should be 60 cm deep. A fast growing, economical tree should be planted where the earth has been refilled at a distance of 5 m from each other. Grasses and other cash crops can grow under the tree. In each hectare of slope land an estimated 750 m long ditches can be constructed.

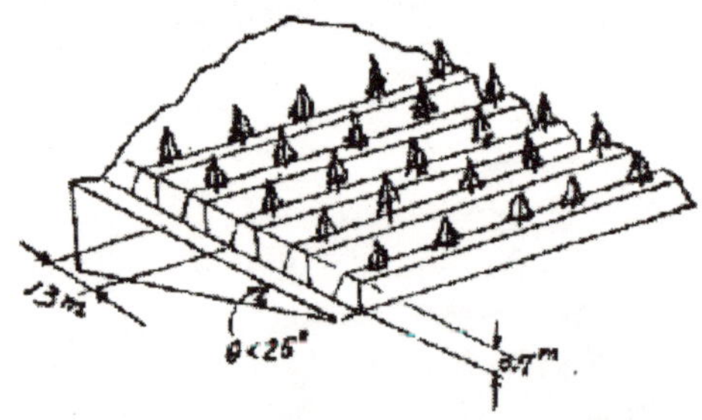

Advantages of the Intermittent Slope Level Ditch

Effectively controls soil and water loss: An area with 750 m/ha of constructed ditches can hold over 900 m^3 soil and water each time. Even a 100 mm rain can not cause any soil and water loss. The slope gradually turned into bench terraces with the natural scouring and yearly land preparation and maintenance.

Improve forest and grass growth: The refilled surface soil brings the fertility to normal top soil level, stores water and conserves soil. This raises the survival rate and growth rate of trees and grasses.

Multi-functions: The constructed ditches are not only favorable for the presently growing crops but also for the preparation of future transformation of slope land into forestland or grassland.

Low labour and other costs: On an average, each hectare of sloping land needs only less than 150 labour days and 75 Yuan (9 U.S. dollars) for tree seedling or grass seeds. Also the technique is very simple for the farmers.

The limitations of above ditches are that they needs lot of labour, consume much land and can reduce productivity in the first 2-3 years.

Southern China Farming Technology for Soil and Water Conservation

Some of the agronomic and cultural practices developed by the people in Southern China are given here.

Contour Cultivation on Terraces

Contour cultivation on terraces is practiced on a large scale for soil and water conservation. It has the capacity to retard runoff, increase infiltration of rainfall and conserve soil and water.

Ditch and Ridge Tillage

This is a technology used in mountainous areas of south China. People plough the sloping lands into contour ditches or into ridged farmland. This increases surface roughness to retard and store runoff, which benefits the growth of crops. The main patterns used in south China are given below.

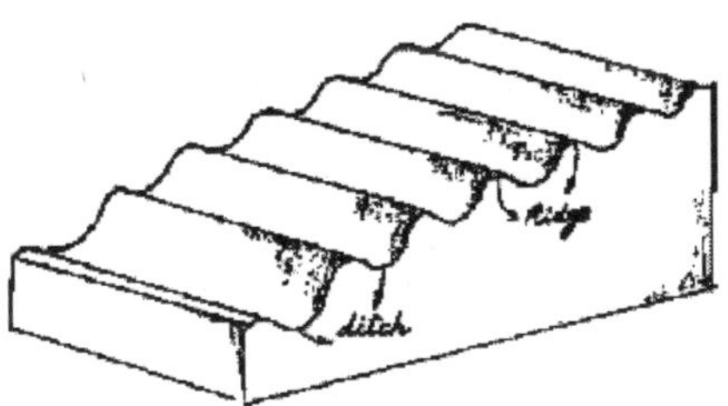

Ditch Tillage for Preserving Soil Fertility

The sloping land is excavated into ditches with width and depth along the contour as required for the row spacing of crops or trees. The top soil is put back with base fertilizer into the ditch.

Contour Ridge Tillage: In this type of tillage a ditch or ridge along sloping land contour is constructed to shorten the slope length and change the direction of runoff flow for the purpose of storing water, preventing scouring and combating drought and soil erosion.

Pit Planting

This technology is widely used in the steep and arid sloping land. Pits of needed size are excavated along contour. The excavated top soil is disposed on the upper side of the slope and kept for refilling. More sub-surface soil is excavated and disposed on the lower side of the slope to form a half-circle ridge for storing water and preventing soil erosion. Then, the stored top soil on the upper side is refilled in the excavated pits along with base fertilizers. The excavated pits between upper and lower part are staggered in order to shorten slope runoff line, increase capacity for storing water and enough space for crop growth.

Semi-dry Rice Farming System

In this system the large paddy fields are changed into ridged and ditched fields to store water in the ditches and transplant rice seedling on the ridges. The advantage of this system is that soil erosion is reduced effectively and the yield of paddy rice increased.

Multiple Crop Rotation

Three crops in a year are grown in south China. It is also called the multiple crop rotation system. For instance, wheat-maize-sweet potato or wheat-yellow bean-sweet potato are grown. It has the advantage of conserving soil and water and increasing crop yield. It makes full use of climatic resources, increases the area for photosynthesis, and prolongs the duration of surface cover.

Minimum Tillage and Mulching System

The characteristic of this system is that the farmland is not ploughed after harvesting the previous crop. Ditches are excavated in rows with fixed spacing for broadcasting. The crop residues between ditches are retained and the previous crop straws or leaves or green manure are used as mulch to cover the newly planted crop. This system has the advantages of reducing soil erosion, reducing loss of organic matter and increasing crop yield.

Contour Strip Inter-cropping System

This system is suitable for sloping lands in south China especially with slopes below 25°.

Grassland strip rotation system: Grassland strips in rotation with crops can effectively control soil erosion, intercept runoff, increase infiltration of soil, improve water quality, and increase crop yield on the sloping lands. The nitrogen fixing grassland rotated with crops will improve soil fertility and structure.

Technique to Pave Gravel on the Sloping Lands for Soil and Water Conservation

Gravel-cover farming is a non-tillage technology used for a long time by people of Gansu Province. Gansu Province is located in the semi-arid and arid region with low rainfall and lack of water resources for irrigation. This is a typical rainfed farming system. In this technique a certain thickness of gravel is paved on the surface soil of the sloping land to store water, reduce evaporation, preserve fertility and reduce soil temperature, sanitation and soil loss.

By this technique, the requirements of water for crop growth can be basically met even under semi-arid and arid climatic conditions without irrigation. Since, the rainfall is generally concentrated in July, August and September, agriculture production is not guaranteed without irrigation. However, by paving gravel in a scattered manner on the surface of sloping lands, soil and water are preserved which can assure a harvest, increase crop yields, increase infiltration, reduce evaporation, temperature and salinization and prevent erosion. The example below illustrates the technique.

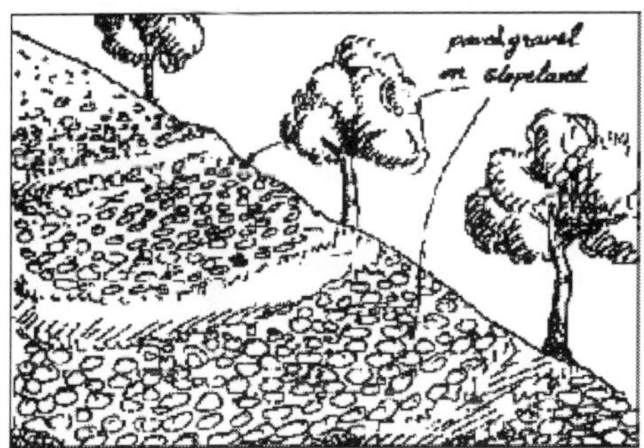

The sloping land is first ploughed once and the soil raked after the fall season. Then 75, 000 kg of organic matter for melon production as well as 150 kg of urea and 300 kg of ammonium calcium in each hectare are applied as base fertilizer. Pits are dug in rows along contour according to the row spacing of melon. The pits should be marked so that the gravel can be put around. Then between rows, some vegetables such as pepper or eggplant can be inter-cropped. Finally, a layer of gravel in 5 cm thickness can be paved on the remaining sloping land. The pits can be used for growing melon during the right season. After the harvest of melon, the second crop, for instance, maize or bean can be grown without tillage.

The gravel-covered farmland can be paved once and used for 15 to 40 years without tillage.

Method of Planting Trees on Terrace Risers

Tree species, which are suitable for planting by inserting them on terrace risers, are used. Strong 1-3 year old seedlings with thickness of 0.5-2.0 cm should be selected and cut into strips of 20-30 cm length. A 3 cm diameter, 60 cm long sharp wood is used to make hole on the terrace risers with a spacing of (0.2m-0.50m) × (0.5m-1.0m). The selected tree stripes are then inserted and the whole is filled back with wet soil. The terrace riser plantation stabilizes terraces and generates income from trees but is limited by rainfall conditions.

Amorpha Fruticosa-a Good Tree Species for Soil and Water Conservation

This species is found in reasons with 1,000 m AMSL. It is widely distributed in northeast China, north China, Guangdong, Yuannan and Guangxi Provinces. It has a well developed root system, is resistant to drought, and requires cold and wet conditions to grow and can survive on saline and infertile soils, seriously eroded gullies in both dry and humid climates. It is of high economic value as it is used for weaving, fodder, fuel wood, manure etc. It retards runoff and conserves soil.

A good Species of Herb for Soil and Water Conservation

Buffalogourd *(Cucurbita foetidissima)* is a kind of perennial herb of the gourd family with perennial roots. It can survive 30-40 years. The plant promises high oil bearing herbal fruits from the very first year to the end and tuber crop underground. It is therefore a kind of economic plant with multiple applications, high adaptability, readily cultivated on barren mountains and floodplains, which are otherwise not arable. An annual intensive farming and fallow cycle is used. Bringing the plant to mountainous areas in north China, which is extremely dry and sparsely vegetated, may be rewarding in conservation of soil and water, while at the same time increasing the income of local people by the sale of herb.

Method of Building Level Strips of Land by Contour Plowing

Sloping lands can be changed in level stripes by contour plowing in about 5 year period. The ridges formed by this method in between two strips can be further reinforced by planting bushes, fodder grass or medicinal plants on them. This adds to more economic benefits. For Inner Mongolia region to develop level strips in the sloping lands below 10° contour plow the lands. On the lower side, the height can increase by 20-24 cm and on the higher side the height can decrease by 20-24 cm. The slope becomes gradual after plowing several times. Experience shows that a 7 m wide belt needs to be contour ploughed 3-4 times. A 10 m level land needs to be ploughed 10 times

Covering Crops with Coloured Plastic Film for Increasing Agriculture Production

Crops like rice seedling and carrot grow faster if covered by a red plastic film. Red colour strengthens the absorption capacity of the leaves and makes the seedling grow healthy which in turn improves the harvest. If a yellow plastic film is used to cover bean, lettuce, celery and cucumber, they will grow well and increase the yield by 50-100%. Farmers can get good harvest and prolong the marketing supply period to get economic benefit. If the temperature gradient of the green house is to be adjusted, silver coloured film should be used.

Good Way to Preserve Agro-plastic Film

There are 3 indigenous methods to preserve and save agro-plastic film

1. Put the clean plastic film in a plastic bag and bind the mouth of the bag and keep in a cool and damp place. This way the film can be preserved for 2 years.

2. Put the plastic film in a big jar with a weight on it, then add water to the jar until the film is submerged and cover it. This way the film can be preserved for 4 years.

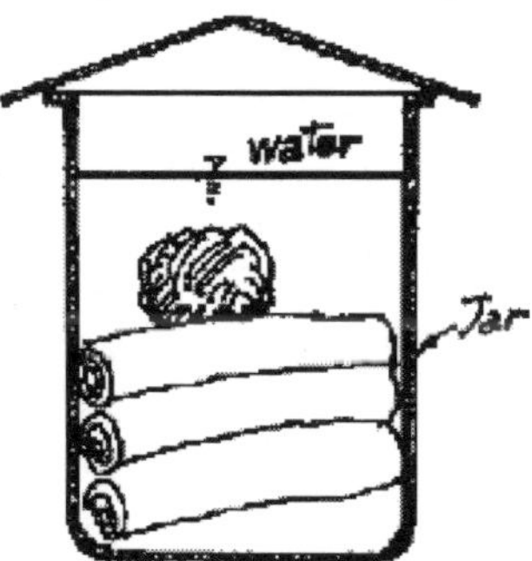

3. Put the clean plastic film in a plastic bag and store it in a 30 cm deep hollow or a cellar. This way the film can be preserved for 3 years.

Chapter 9

Recent Advances in Dryland Agriculture

Indian agriculture is predominantly a rainfed agriculture under which both dry farming and dry land agriculture is included. Dry faring was the earlier concept for which amount of rainfall (less then 500 mm annually) remained the deciding factor for more then 50 years. In modern concept, dry land areas are those where the balance of moisture is always on the deficit side. In other words, annual evapotranspiration exceeds precipitation. In dry land agriculture, there is no consideration of amount of rainfall. It may appear quiet strange to a layman that even those areas which receive 1100 mm or more rainfall annually fall in the category of dry land agriculture under this concept. To be more specific, the average annual rainfall of Varanasi is around 1100 mm and the annual potential evapotranspiration is 1500 mm. thus the average moisture deficit so created comes to 400 mm. this deficit in moisture is bound to affect the crop production under dry land situation ultimately resulting into total or partial failure of the crops. Accordingly the production is either low or extremely uncertain and unstable which are the real problems of dry land in India.

The success of crop production in these areas depends on the amount and distribution of rainfall, as these influences the stored soil moisture and moisture used by crops. The amount of water used by the crop and stored in the soil is governed by the water balance equation: $ET = P-(R+S)$. When the balance of the equation shifts towards right, precipitation (P) is higher then ET, so that there may be water logging or it may even lead to run off (R) and flooding. On the other hand, if the balance shifts to the left, ET becomes higher then the precipitation, resulting in drought in the various severity. Taking the country as a whole, as per meteorological report, severe drought as large area is experienced once in 50 years and partial drought in five years while folds are expected every year in one part of the country or the other, especially during rainy season. In fact the balance of the equation is controlled by the weather, season, crops and cropping pattern.

Out of 14.2 million ha of net sown area in the country, rainfed agriculture is practiced in 95 million ha (67%). Nearly 67 m ha of rainfed area falls in the mean annual precipitation range of 500-1500 mm. The average annual rainfall of the country is 1200 mm amounting to 400 million ha meter of rain water over the country's geographical area (329 m ha). However, the distribution across the country varies from less than 100 mm in extreme arid areas of western Rajasthan to greater than 3600 mm in NE states and 1100 mm from east coast 2500-3000 mm in the west coast. The broad area of the summer monsoon activity extends between 30^0 N to 30^0 S and from 30^0 W to 16.5^0 E. the detail information on rain fall and monsoonal pattern in India has been summarized in the following table:

Table: *Rainfall pattern in fall*

Season/Period	***m ha m***	***Percent***
Winter (Jan-Feb)	12	3
Pre-monsoon (Mar-May)	52	13
Southwest monsoon (Jun-Sept)	296	74
Northeast monsoon (Oct-Dec)	40	10
Total for the year	400	100

Rainfed farming comprises about 91% area of coarse cereals (sorghum, pearl millet, maize and finger millet), 91% pulses (chickpea and pigeon pea), 80% of oilseeds (groundnut, rape seed, mustard and soybeen), and 65% of cotton. Also, about 50% area under rice and 19% area under wheat is rainfed.

During the past 25 years there occurred significant changes in the area and yield of imported crops of rainfed farming areas. The area under coarse cereals decreased by about 10.7 million ha and most of this was under sorghum. The area under oilseeds increased by 9.2 million ha and most of this increase was due to irrigated rapeseed and mustard and soybeen. The total area under pulses and cotton remained constant but more of cotton became irrigated and shifts in the area under occurred from one agro-ecological region to others. Area under chickpea in the northern belt decreased but increase in the central belt. This change occurred due to increase in area under rice-wheat cropping system which displaced chickpea and also pearl millet to a great extent and maize to a small extent.

According to the present concept, there are 128 districts in the country which face the problems of dry land of these 25 districts covering 18 m. ha of net area sown with 10 % irrigation receive 375-750 mm rainfall annually spread over Central Rajasthan, Saurashtra region of Gujarat and rain shadow region of Western Ghats in Maharashtra and Karnataka. Twelve districts have irrigation covering 30-50% of the cropped area and do not pose serious problems. The remaining 91 districts covering mainly *Madhya Pradesh, Gujarat, Maharashtra, Andhra Pradesh, Karnataka, Uttar Pradesh, parts of Haryana, Tamil Nadu* etc., represent typical dry land area. The total net sown area in these districts is estimated to be 42 million hectares of which 5 m ha are irrigated. Rainfall in these districts varies from 375 to 1125 mm. therefore, more and more efforts are to be made for enhanced and stable production in these areas so that the recurring droughts do not stand in the way of meeting the growing food demands.

It is not that no attention has been paid in the country towards the development of dryland farming. Efforts were made right from 1923 to improve crop yields with the establishment of a research projects at Manjari in Maharashtra and later at Solapur, Bijapur, Raichur and Hagari in Deccan and Rohtak (Haryana) in the north. An All India Coordinated Research Project for Dry land Agriculture was launched by ICAR in 1970 in collaboration with Government of Canada and later Central Research Institute for Dry land Agriculture (CRIDA) was established in 1985 at Hyderabad. These projects generated technology, which, if followed, can bring marked improvement in cropping intensity, productivity and stability in production.

In dry land agriculture, scarcity of water is the main problem. Apart from the low and erratic behaviour of rainfall, high evaporative demand and limited water holding capacity of the soil constitute the principle constraint in the crop production in dry land area. Yield fluctuations are high mainly due to vagaries of weather, often much behind the risk bearing capacity of the farmers. It is surprising to a layman that even humid areas with 2000 mm of annual rainfall not only suffer from moisture stress, but also face drinking water scarcity. Monsoon starts in the month of June and ends in last week of September or sometimes in the first week of October. Most of the rainfall is received during this period. With undulating topography and low moisture retention capacity of the soil, major portion of the rain water is lost through runoff, causing erosion and adding to the water logging of low lying areas. After the rain stops, very little moisture is left in the profile to support plant growth and grain production. In dry land area deficiency and uncertainty in rainfall of high intensity causes excessive loss of soil through erosion which leaves the soil infertile. Owing to erratic behaviour and improper distribution of rainfall, agriculture is risky, farmers lack resources, tools become inefficient and ultimately productivity is low.

Vertisoles have high clay content and high moisture retention capacity. Owing to its swelling and shrinking characteristics, permeability is low and hence the rate of infilteration of water is minimum. This causes more surface and high soil loss from the top layer owing to surface erosion. It is estimated that 68.5 tones/ha per year soil is lost from *vertisoles.* Due to high clay content it develops cracks during Rabi season at flowering stage of crops.

Alfisols are, by and large, light textured soils which have low moisture holding capacity but high water intake. The rain water falling in such areas gets soaked up and saturates the profile. The soil water percolation is more and therefore, is lost for crop use. Owing to faster intake of water in the profile the surface runoff is limited and soil loss from erosion is low (3.05 t/ha/year). Soil crusting is a common problem in low rainfall areas.

Entisols are generally loamy sand or sandy loam. Depth in these soils is not a constraint. These soils have very low clay content and hold water up to 200 mm per meter of soil profile. Its nutrient holding capacity is poor. In low rainfall areas monsoon cropping is practiced and in high rainfall areas double cropping is possible.

Submontane soils are medium in texture and depth is medium to deep as well as moderate in clay content. Moisture retention capacity is high (300 mm/m. profile). These soils are poor in nitrogen but in other nutrients. Phosphorous may be limiting in high production system. Due to high rainfall double cropping is possible in these soils.

Sierozems are extremely light soils, effectively depth being influenced by the $CaCo_3$ concentration in soil profile. Its moisture holding capacity is low (150 mm water.m). Sierozemic soils are low in nitrogen and sometimes inadequate in phosphorous. Subsoil salinity is common. These soils are mostly monsoon cropped, except in deep sandy loams where post-monsoon cropping is also possible. Crusting is very frequent.

Improved Dryland Technology

The improved techniques and practices, which have so far been generated and recommended for achieving the objective of increased and stable crop production in dryland areas, have been summarized in following lines.

Crop Planning

The farmers of the dry land areas, prior to the development of dry land techniques, were growing a crop either on rainwater in kharif or on conserved soil moisture during the winter. The crop varieties grown when moisture is sufficiently available. Such varieties have low genetic potential for yield. Selecting suitable crops and varieties capable of maturing with in actual rainfall periods will not only help in enhancing production of a single crop but in intensifying the cropping intensity. Many criteria have been laid out for selecting a crop variety for drylands. The capacity to produce a fairly good yield under limited soil moisture conditions is the most desirable criteria.

The duration of kharif crops/varieties should not normally exceed the number of rainy days. In other words, crop varieties for dryland areas should be of short duration, through resistant tolerant and high yielding which can be harvested with in rainfall periods and have sufficient residual moisture in soil profile for post-monsoon cropping.

Under dry land agriculture determination of length of growing period (LPG) i.e., moisture availability of a given soil type, provides better index than total rainfall based crop planning. LPG is defined as the period when the moisture and temperature regious are suitable for crop growth and the period is determined by the FAO method (1983). The LPG is computed as the sum of the period when P is more than 0.5 PET plus time taken to utilize stored soil moisture (assured 100 mm) after P falls short of PET. For example 'Nagpur' and Ratnagiri in Maharashtra receive mean annual rainfall of 1120 mm and 2500 mm, respectively but LPG determination indicates that both the places have LPG of 210 days in deep black soils. Therefore, both the places are suitable for single long duration on a short duration crop with a relay rabi crop.

Planning for Aberrant Weather

Dryland agriculture is subject to high variability in areas sown, yields and output. These variations are the results of aberrations in weather conditions, especially rainfall. Delays in normal monsoonal pattern causes problems of timing and the organization of preparatory tillage and other initial activities for commencing cultivation processes for the season. Such monsoonal delays have repercussions on the programme of activities for the entire agricultural year. Even after the onset of monsoon and the commencement of planting, there may be monsoonal withdrawal causing moisture stress on plants and creating difficulties in the adoption and timing of approval cultural practices ultimately causing reactions in yields and outputs. Some crops are highly susceptible to such mid-season variations in moisture availability such as at the flowering stage in rice. Major crops like rice and maize get seriously affected if monsoonal rains cease early.

The need for modifying and introducing and introducing new technology for increasing and sustaining yield in dry land areas can hardly be overemphasized. Equally urgent is the need to decelerate and ultimately eliminate the process of damage to agricultural assets which are proceeding unbated in dry land areas. Erratic rainfall results in fluctuating production. This in turn leads to frequent scarcities, like the ones experienced in Indonesia and Vietnam in 1977 which created severe food shortages. Droughts in China in 1972, 1974 and 1985 brought depression of food grain production by up to about 25 million tones. Frequent droughts in India during 1966, 1968, 1972, 1974, 1979, 1982 and 1987 seriously affected the food and fodder production in the country. Hence, it is necessary to understand the distribution of South-West monsoon within the season to determine the extent to which the crop productions are likely to be affected by the vagaries of monsoon.

Several attempts have been made to understand the behaviour of South-West monsoon rainfall in different agro-climatic regions on the basis of historical rainfall records. These studies have brought out that (i) there is large variation in dates of commencement of South-West monsoon from year to year in different parts of the country, (ii) the monsoon rainfall is of sequential nature with long dry spells or breaks extends sometimes to the period of even one month or more, (iii) there is large year to year variation in dates of withdrawal of South-West monsoon, (iv) there is variation in quantum of rainfall received from year to year and (v) high intensity rainfall occurs in association with movement of cyclones or depression resulting in sizeable loss of rainwater through run-off and deep drainage.

Thus, crop production in dry lands fluctuates widely from year to year due to vagaries of weather. An aberrant weather can be categorized under three heads i.e. (i) delayed onset of monsoon, (ii) long gaps or breaks in rainfall, and (iii) early stoppage of rains towards the end of monsoon season. Therefore, to mitigate such weather situations, farmers should make some changes in normal cropping schedule for getting some production in place of total crop failure.

Crop Substitution

Alternate crop strategies have been worked out for important regions of the country for *vertisols, alfisols, entisols, submontane* and *sierozemic* soils. Strategy has also been evolved for normal onset of rains, breaks in rains, early withdrawal, its uneven distribution; through selection of uneven crops/varieties which are inefficient utilize of the soil moisture, less responsive to production input and potentially low producers should be substituted by more efficient ones. Appropriate crops, suiting varying rainfall situations, have been identified for most of the dry land regions of India.

Crops which do not under normal rainfall years may not do so under abnormal years. Studies conducted in agro climatic conditions of Varanasi (eastern U.P.) revealed that under normal monsoon crops like short duration upland rice, maize, pearl millet, blackgram, greengram, sesame, pigeon pea etc. should be taken up on the basis of needs. These crops should be followed by chickpea, lentil, barley, mustard, safflower, linseed etc. on residual moisture during winter season. If monsoon sets in as late as second week of July, short duration upland rice (variety-NDR-97 and NDR-118) may be included in place of Akashi and Cauvery is recommended.

If the rains are delayed beyond the period but start somewhere in last week of July or first week of August and growing season is reduced to 60-7-days, then cultivation of hybrid pearl millet (NHB 3-4, B.J. 104), blackgram (Type 9), greengram (var-Jagriti and Jyoti) may be included in pace of T-44 and k-851 etc. should be grown. Yet another alternative could be to harvest a fodder of either pearl millet, maize, sorghum or a mixture of cowpea, blackgram and one of the above fodder crops.

In case monsoon rains stop early towards the end of season, normal sowing of short duration upland rice, blackgarm and sesame may be taken up.

If the rain stops very early, i.e. by the end of August or first week of September, only fodder crops or grain legumes could be harvested. Depending upon the soil moisture condition, relay sowing of crops like chickpea, lentil, mustard, linseed and barley could be done in rabi season.

Table: *Traditional and Alternate Efficient crops in Different Dryland Regions of India*

S.N.	***Region***	***Traditional crop***	***Alternate efficient crop***
1.	Deccan Rabi season	Cotton, wheat	safflower
2.	Malwa Plateau	wheat	Safflower, Chick pea
3.	Uplands of Bihar Plateau and Orissa	Rice	Ragi, Black gram, Groundnut
4.	South-easy Rajasthan	Maize	Sorghum
5.	North Madhya Pradesh	Maize	Soybean
6.	Eastern UP	Kalitur	Chick pea
7.	Sierozems of	Wheat Northwest India	Mustard, Taramira (Eruca sativa)

During the recent drought, it was found that farmers in Karnataka, Andhra Pradesh and Maharashtra who went in for sunflower cultivation were in gainers. Sunflower succeeded where other crops failed. In other dry land regions, alternative efficient crops can profitably substitute the traditional ones (Table 2).

Table: *Relative Yield of Traditional and Efficient Crops in Dryland Areas*

Region	*Traditional*	*Yield (q/ha)*	*Efficient crops*	*Yield (q/ha)*
Bellary	Cotton	2.0	Sorghum	26.7
Varanasi	Wheat	8.6	Chickpea	28.5
Ranchi	Upland Rice	28.8	Maize	33.6
Indore	Green gram, Wheat	11.8	Soybean	33.3
		11.0	Safflower	24.2
Agra	Wheat	10.3	Mustard	20.4
Hisar	Wheat	3.0	Taramira	16.0
Udaipur	Maize	18.0	Hybrid sorghum	29.0

Dry land research has remained confined to important traditional crops such as sorghum, millet, pulse and oilseeds and has not explored the possibility of growing non-traditional crops such as dye-providing crops {e.g. Henna (Lawsonia inermis: mehadi) and jaffra(Bixa ovellana) species (0e.g. cumin), and medicinal value crops (e.g. eitronella, lemon grass, senna and isabgol)}.

These crops need to find an important place in research aagenda of dry land farming. Time has come for the relevant researchers to plan a joint integrated research programme for maximizing the profitability, productivity and sustainability of learning systems of rainfed areas. Sericulture offers great promise in rainfed farming strategy, particularly of the watershed approach in peninsular India.

Efficient Cropping System

Besides putting various measures to increase the productivity levels of dry land crops, efforts would also be needed to increase the cropping intensity in dry land areas which was generally 100%, implying that a single crop was taken during the year. Cropping intensities of these areas could be increased by practice of inter cropping and multi cropping (sequential) by way of more efficient utilization of resources. The cropping intensity would depend on the length of growing season which in turn depends on rainfall pattern and the soil moisture storage capacity of the soil. For example in Indore region, receiving 1000 mm annual rainfall, only single crop can be taken on shallow soils, inter cropping in medium depth soils and double cropping on deep soils. Similar crop combinations have been identified for different regions of the country. In dry land of Varanasi region upland rice-chickpea/lentil sequence can be practices with advantage. Inter cropping of vegetables with grain crops was pursued vigorously in some centres such as Varanasi and Phulbani. At both the palces long duration pigeon pea was inter cropped with vehetables such as okra, radish and chilli. Such inter cropping systems would be very useful to get maximum returns from rainfall agriculture. Even at solapur, leafy vegetables and some short duration beans were grown as intercrops during the rainy season.

Fertilizer Use

Soils of dry lands in the country are not only thirsty but hungry also because these soils are severely eroded horizontly as well as vertically. Whenever efforts are made towards bunding and levelling of the fields in dry land areas, it is the surface soil which is removed. The resultant effect is that the fields are rendered shallow in depth and completely deprived of plant nutrients, particularly nitrogen, phosphorus and potassium. It is, therefore, necessary to apply all the three major nutrients in adequate amounts. Since soil moisture is limiting in dry lands, the availability of nutrients becomes limited, attempt should always be made to apply fertilizers in furrows below the seed. If seed-cum-fertilizer drills drawn by bullocks or tractors are available, this very objective can be fulfilled.

There has been belief among the farmers of dry land areas that use of fertilizer increases the chances of crop failure but recent findings have shown that the use of fertilizer is not only helpful in providing nutrients to crop but also helpful in efficient use of profile soil moisture. If dry land farmers are shown such results, they will be convinced to use ore and more fertilizers. Studies on the management of legumes in crop sequences for their residual effect indicated that in alluvial soils an advantage of 25-30 kg N/ha could be obtained in barley or mustard grown after black gram or green gram. Another possibility for nitrogen management in cropping system is to use legumes as green manures either at flowering stage or after one picking. Studies conducted at Varanasi centres clearly showed that general yield levels of barley and mustard were greater when legumes raised in the previous season was incorporated I soil after first picking as compared to that harvested at normal maturity.

In dry land areas, a proper mixing of organic and inorganic would be desirable. Organics have low nutrient content, but help to improve the moisture holding capacity of soils. In addition to yield advantage, nutrients like potassium help to increase drought tolerance by affecting plant-soil relationship. Transpiration losses are reduced and productivity per unit water increases.

Table: *Effect of N-levels on Yield and Moisture Use Efficiency (MUE) of Barley and Wheat (Varanasi Centres)*

Nitrogen levels (kg/ha)	*Grain yield (q/ha)*	*Total moisture use (mm)*	*MUE (kg/mm)*
Barley			
0	14.05	133.7	10.5
30	20.45	136.3	15.0
60	30.00	142.3	21.0
90	37.20	141.6	26.3
Wheat			
0	9.55	145.5	6.6
30	13.55	144.4	9.3
60	18.35	153.6	119
90	24.15	155.1	13.6

Table: *Nitrogen Economy to Legume-Cereal System (4 years average)*

Nitrogen level (kg/ha)	***Incorporated***	***Crop yield (q/ha)***	***Unicorporated***
Green Gram	**1.89**		**2.23**
Barley			
0	16.98		13.65
30	21.30		18.64
60	24.43		21.84
90	27.27		25.20

Rain Water Management

Efficient management of rain water can boost agricultural production from dry lands. The broad bed and furrow system of the International Crop Research Institute for the Semi Arid Tropics (ICRISAT) for managing rain water in vertisols made it possible to increase crop yields four to five times as compared to normal practice. However, this method could not be adopted widely by the farmers in India because it is costly and labour intensive. The vertical mulching developed at Bellary centres increases the infiltration of water in soil profiles and improves in situ moisture conservation. The scope for managing profile moisture is limited in alfisols but the surface run off in such soils can be reduced by ridge-and-furrow technique. Alternatively, application of compost and farm yard manure as well as raising legumes will add the organic matter to the soil and increase the water holding capacity.

The winter which is not retained by the soil flows out as surface runoff. The run-off-recycling holds immense prospects in deep black soils where the seepage losses are very much less. This runoff water, if not permitted to drain out safely, causes erosion. Therefore, safe disposal of excess water from the field drains to the disposal system should be planned properly. This excess runoff water can also be harvested in storing dug out ponds and recycled to donor area in the event of severe moisture stress during rainy season or for raising crops during the winter.

Water-shed Approach for Resource Improvement and Utilization

Watershed management is a holistic approach arrived at optimizing the use of lad, water and vegetation in an area and thus, providing solution to alleviate drought, moderate folds, prevent soil erosion, improve water availability and increase fuel, fodder and agricultural production on a sustained basis. On the basis of the experiences of ICAR Operational research Projects, which attracted the attention of our farmers, State departments, administrators and scientists, 47 model watersheds were established during the year 1983 for development, jointly by the Ministry of Agriculture, ICAR and various State Government Department and Agricultural Universities, in 16 states and then the Department of Agriculture and Cooperation launched the National Watershed Development Project for Rainfall Areas (NWDPRA) covering almost the same states. Out of these 47 model watersheds, the Central research Institute for dryland Agriculture (CRIDA). Hyderabad has been entrusted with 30 watersheds. These activities were in micro and mini-watersheds covering 500-2000 ha. Major components in these model watersheds are: (i) Improvement of water resources, (ii) In situ soil and water conservation: rain water harvesting for safe disposal of surface runoff, (iii) increase in cropping intensity and (iv) alternate land use system for efficient use of lands as per land capability to provide stability in productivity.

The model watersheds in operation have provided a fruitful experience of how development can lead to all round improvement in food and fodder production, economic condition of the farmers. Sakho-majori model, where creation of eater source worked as a catalyst and triggered the development process can be repeated under similar situations. Similar experiences have been gained a Tejpura (Jhansi), Ariel (Bareilly District) and Tejpura watersheds which have been awarded the First and Second Prizes respectively by the President of India on 14-11-1988 based on the recommendation of National Productivity Council.

Alternate Land use System

All dry lands are not suitable for crop production. Some lands may be suitable for range/pasture management, while others for tree farming, ley farming, dry land horticulture, agro-forestry systems including alley cropping. All these systems which are alternatives to crop production are called as alternate land use systems. This system not only helps in generating much needed off-season employment in mono crop dry land but also minimizes risk, utilizes off season rains which may otherwise go waste as runoff, prevents degradation of soils and restores balance in the ecosystem.

Crop production may be disastrous in the years of drought, whereas drought resistant grasses and trees could be remunerative. Scientists of dry land have developed many alternate land use systems which may suit different agro ecological situations. These are alley cropping, agri-horticultural system and silvi-pastoral systems which utilize the resources in better way for increased and stabilized production from dry lands. Alley Cropping: for imparting stability and providing sustainability to the farming system, a tree-cum-crop system will be one most appropriate for such situations. One such system called 'alley cropping'-a version of agro-forestry system, could meet the multiple requirements of food, fodder, fuel, fertilizer etc. Alley cropping is a system in which food crops are grown in alleys formed by hedge rows of trees or shrubs. The essential feature of the system is that hedge rows are cut back at planting and kept pruned during cropping to prevent shading and to reduce competition with food crops.

For example, fast growing leguminous trees such as *Leucaena leucocephala* or *liliricidia spp.* are planted in rows. During the cropping season, trees are lopped at about 0.5 metre height. These loppings are used as much to reduce moisture loss and improve the nutrient status of soil. Arable crops like maize, rice, pearl millet, legumes, oilseeds etc. are planted in the alleys formed by the two rows of threes. This is also known as agri-silvi culture. Alley cropping is also a form of conservation farming which enhances soil fertility and prevents erosion.

One very strong argument in favour of alley cropping is its ability to produce usable material even in years of severe drought. At Rajkot in 1985, rainfall received during the season was only 30% of the normal. There was total failure of grain production of the three legume crops tried in the system. In sole crop plots production was limited to 5.0 q/ha to 17.0 q/ha of green fodder. However, in alley cropped plots, *Leucaena* hedge-rows produced over 50.0 q/ha of green fodder. Agri-horticultural system: Agri-horticultural system palys an important role in dry land areas, especially in semi-arid regions where production of annual crops is not only low but also highly unstable. Fruit trees if suitably integrated in dryland farming system could add significantly to overall agricultural production including food, fuel and fodder, conservation of soil and water and stability in production and income. Dry land fruit trees being deep rooted and hardy, can better tolerated monsoonal aberrations than short duration seasonal crops. Hence, in drought year when annual crops usually fail or their production is highly depressed, fruit trees species yield considerable food, fodder and fuel.

A suitable example of agri-horti-system is growing of cow pea/green gram/horse gram in inter space of *ber (Zizyphus mauritiaria)* at 6 x 6 m spacing at Hyderabad. Phalsa (*Grewia asiatica*) may be planted in between two ber plants in a row with a view to intensify the system. A well managed dry land orchard of *ber* should give 50 kg fruits per tree/ year. There should be 250 plants/ha for optimized production. The grow income would touch around Rs. 50,000/ha (250 x 50 x 4), assuming that one kg *ber* fetches Rs. 4. One could get an additional income of Rs. 800-Rs. 1000 from green gram/cow pea (2.5-3.0 q/ha).

Silvi-Pastoral System: This system is suited to marginal dry lands and is most preferable where the fodder shortages are experienced frequently. The system essentially consists of a top feed tree species carrying grasses on legumes (preferable perennial) as understorey crops. Dry land farmers having larger holdings and keeping a land follow for a longer period for one reason on the other, should go in for this system which could provide both fodder and fuel. In a survey carried out in Andhra Pradesh, Karnataka and Maharashtra by CRIDA scientists, it was revealed that after food it is the fodder which is of paramount importance for sustaining animal wealth in rural areas. In years to come, fuel will assume greater importance.

In August, 1981 *Leucaena leucocephalla* was planted in contour trenches 7.5 m apart, the plant to plant spacing being maintained at 2.0 m at CRIDA. Four strips at upper reaches of plot (2% slope) were put under *Cenchrus ciliaris,* while lower four strips were seeded with *Stylosanthes hamata*. The system has come up very well.

Efficient Implements

In order to take full advantage of annual precipitation in dry land agriculture, higher doses of energy input is essential. Farmers in dry lands have been using traditional and outdated farm equipments which not only perform poorly but also demand a lot of energy and time and post-harvest operations. Farm implements can help to conserve as much rain water in situ as possible and to harvest rain water. Shallow off season tillage with pre-monsoon showers ensures better moisture conservation and lesser weed intensity. It has resulted in 20% yield increase in sorghum in Andhra Pradesh. Deep tillage helps in increasing water in soils having textural profiles and hard pan. This has resulted in 10% yield increase in sorghum and 9% yield increase in case of caster. For in-situ moisture conservation, land has to be opened so that it can cause hurdle to flow of rain water. Tillage machines of appropriate size and type matching the power sources need to be used. Location specific seeders have been developed for dry land areas and these have shown good prospects and promise. A feature of these machines is that the seeds and fertilizers are placed in the moist zone of the soil resulting in a high percentage of seed germination and good crop vigour. In deciding farm mechanization in dry land areas, where farmers are generally poor, and their socio-economic condition should always be kept in mind. The foregoing discussions show that technology of crop production in dry land areas have been generated to a great extent. What is important now is to view it in socio-economic context of the farmers. Once the technology is adopted by the farmers, the contribution of dry land areas to the total production can be sizably improved and the living standards of the farmers of these areas can be improved. This has been clearly shown in selected watershed areas and what is needed is to have more watersheds identified, proper technology to be developed and implemented.

Unlocking the Hidden Potential of Dryland Agriculture

IF government statistics are to be believed, the Indian economy never had it so good. For the past several years, India has been one of the fastest growing economies in the world. The rate of growth of GDP recorded during the first three years of the tenth plan (7%) is the highest since planning began in 1951. Other macroeconomic indicators like the rate of saving, current account balances and inflation rates seem to indicate the economy is in good health.

Against this background, the story of Indian agriculture, in particular dryland agriculture, appears as an annoying twist in the plot. Available data show that 1990-2000 was not a happy decade for Indian agriculture. The overall growth rate of crop production during this decade was nearly half of what it was during the 1980s. The output of coarse cereals, pulses and oilseeds (covering about 45% of total cropped area and grown mostly in the drylands) fell during the 1990s and the rate of growth of their yields decelerated considerably. All this re-emphasise the fact that the gap between irrigated and dryland agriculture has steadily widened, with the productivity of the latter being less than half of the former. This has been a direct and predictable consequence of the strategy adopted in the mid-1960s, whereby massive investment flowed to the already well-endowed regions and farmers of the country, leaving aside the poorer and less endowed regions.

Dryland agriculture emerges as the biggest drag on the growth of the economy. Indeed, the dominant strand of thinking among our policy-makers treats the drylands as a hopelessly lost bet. However, we should caution ourselves against such a hasty conclusion. Even at their low productivity levels, the quantitative significance of dryland agriculture is by no means small.

It accounts for 53% of total cropped area, 48% of the area under food crops and 68% under non-food crops. In terms of production, drylands account for nearly 80% of the output of coarse cereals, 50% of maize, 65% of chickpea and pigeonpea, 81% of groundnut and 88% of soyabean. Half the output of cotton in the country is from the dry districts (Shah et al., 1998). Given its large size and extremely low productivity levels, a unit rise in productivity in this sector is likely to have the largest impact on aggregate crop productivity. There is clear evidence that the yield potential of dryland varieties is much higher than what has been achieved on the farm. It is more appropriate to view the drylands as a source for future growth, a hidden potential waiting to be unlocked.

Realising the potential of drylands is by no means a simple task. Spread over nearly half of the country, the drylands cover cold arid regions, hot deserts, hilly and undulating uplands, forest areas, plateaus, ravines and coastal and non-coastal saline areas. They are the home to 43% of our population. Water availability, soil conditions and the length of the growing season show wide variations here. Nine states (Rajasthan, Madhya Pradesh, Maharashtra, Gujarat, Chhatisgarh, Jharkhand, Andhra Pradesh, Karnataka and Tamil Nadu) account for over 80% of the drylands. Annual rainfall in the drylands varies from less than 150 mm to 1600 mm. Soils vary from shallow skeletal soils of the deserts to medium to deep black soils.

The list of prominent dryland regions brings out their diversity:

1. Western Himalayas: Cold arid region with rainfall <150 mm and shallow skeletal soils;
2. Western Rajasthan, Kutch and northern part of Kathiawar peninsula: Hot arid region with rainfall <300 mm, desert and saline soils;
3. Rajasthan uplands (Aravallis) and Chambal districts of Madhya Pradesh: Semi-arid region with alluvium-derived soils and extensive land degradation leading to ravines;
4. Central Highlands, including Gujarat plains and western Madhya Pradesh (Malwa): Semi-arid region with rainfall of 500-1000 mm, medium and deep black soils;
5. Deccan Plateau, including Maharashtra and northern Karnataka: Semi-arid region with rainfall of 600-1000 mm, red and black soils;
6. Interior Andhra Pradesh (Telangana): Semi-arid region with 600 Â– 1000 mm rainfall, red and black soils;

7. Tamil Nadu Uplands and western Karnataka, semi-arid region with red loamy soils;
8. Sub-humid Eastern Plateau (Chhatisgarh), with rainfall of 1000 to 1600, red and yellow soils;
9. Sub-humid Eastern Chhotanagpur Plateau including Jharkhand, western Orissa and northern Andhra Pradesh, with rainfall of 1200 to 1600, red and lateritic soils.

The range and diversity of the drylands presents a qualitatively complex set of constraints which needs a more delicate mode of handling. We must give up the one-size-fits-all approach and focus on fine-tuning and matching our interventions to the subtle variations in local contexts. At the outset, we must realize the crucial role the state has to play in a positive reshaping of the dryland landscape. In the current context of liberalization and dominance of anti-state sentiments (shared, paradoxically, by the staunchest of neo-liberals as well as the purest among radicals), this point needs strong emphasis.

In particular, the key role of kickstarting the growth process in the forsaken drylands has to be played by public investment. The drylands are caught in a low-level equilibrium trap. Public investment is required to enable them to break out of this trap. The endemic process of natural resource degradation needs to be checked. Public investment in drylands has to be substantial, multi-directional and sustained over a long period of time. What we see at the national level is a clear decline in public investment in agriculture since the mid-1980s. This trend needs correction and the state has to step up its investment in the drylands substantially.

The precise forms such investment should take vary widely from region to region. Rather than being imposed from above, they must organically develop through an active exchange with the local context and its problem typology. There are examples galore to show how imposed, inappropriate interventions cause havoc. The desert region of western Rajasthan is one of the hottest regions of the country receiving scanty rainfall (100 to 300 mm) every year. Soils are dry during most part of the year. There have been several major attempts to 'green' the desert landscape. The most celebrated of them is the Indira Gandhi Nahar Project (IGNP), a gigantic water transfer project of diverting water to the desert.

Today, excessive irrigation and poor drainage planning has led to extensive waterlogging and land degradation due to sodicity-alkalinity. There are reports of hard pan formation beneath the soil layer. The severe ecological damage of this project that entailed an attempt to extend the green revolution to the desert is a clear pointer to the limits of human action in mastering nature. Yet, mega-schemes like extending the IGNP canal or lifting and carrying water from Narmada are axes along which solutions to the water problem of this parched desert landscape are envisaged.

Interestingly, the desert has a long tradition of local water harvesting for drinking water purposes and limited amount of agricultural use. Public investment could strengthen these. *Tanka* (or underground cisterns located in the middle of a large circular area which is its artificial catchment), *nadi* (community ponds usually constructed in inter-dunal spaces where there is some surface water run-off because of underlying geological conditions), *dighi* (small square or semi-circular, step-reservoir) and *khadin* (embankment built across slopes in agricultural fields) are some of the traditional water storage structures of the desert. Similarly, there are groundwater harvest structures as well, such as *baoris*, *kund*, *kuiya* etc. The desert has developed some locally appropriate agro-forestry systems and traditional systems of addressing the problem of wind erosion and shifting sand dunes. A clear understanding of the scope and limits of such systems is necessary before we impose our external solutions through ill-conceived mega schemes.

Many other parts of the drylands also have traditional water management systems, such as tank irrigation in semi-arid South India and southeastern Rajasthan and the *ahar-pyne* system of South Bihar plains. Without romanticizing 'tradition', we can say that these systems show minute awareness of variations in topography, soils, rainfall and crop needs. However, these have been designed for much smaller populations and their institutional support systems have decayed over time. Rejuvenating and strengthening these must be an important plank of policy intervention.

The experience of borewell irrigation technology in hard rock regions of western India shows that there can be no miracle solutions to the dryland challenge. The Malwa plateau in western Madhya Pradesh is a case in point. The semi-arid Malwa plateau has predominantly hard rock geology, being underlain by Deccan Trap basalts. The natural rate of recharge of groundwater in hard rock regions is very low. Hard rock aquifers can often be fairly large reservoirs of stored water (accumulated over several thousand years). However, once they dry up, they cannot easily be replenished. Increasing the depth of groundwater extraction creates a very real danger of groundwater mining. This means that great caution needs to be exercised in the extraction of groundwater here. Such a careful approach never seems to have informed the groundwater economy of Malwa.

Since the introduction of the green revolution package, aided by bore-well technology since the 1980s, farmers in the Malwa region have been on a hot chase of extracting as much groundwater as they can. Subsidised electricity has accelerated the process. At present, uncontrolled extraction has given rise to a severe, man-made crisis of groundwater here. 60% blocks in the region are classified as dark or overexploited, meaning that their levels of groundwater extraction are unsustainable. Falling water levels are reported from everywhere and the area is in the throes of a severe drinking water shortage during summer. Yet, groundwater mining through deeper drilling continues unabated.

The priority intervention here is replenishment of highly depleted groundwater aquifers through strategically located surface water storage structures and sub-surface dams. These supply-side interventions should go together with regulation of end-uses of water through community management of groundwater, low cost drip irrigation and diversification of the crop production systems using drought-resistant varieties and crops that require less water. A different package of interventions needs to be adopted for medium-rainfall eastern India (covering Jharkhand, non-coastal Orissa and Chhattisgarh), where the problem is severe underutilisation of available water. These economies are characterised by high subsistence orientation, small size of landholdings, poor irrigation infrastructure and heavy outmigration. Within this region, the tribal pockets are more deprived than others. Varying land slopes, water availability and soil depth define several land situations here. Public investment should support a basket of community-based strategies focused around utilisation of available water resources to improve productivity in each land situation. Detailed grassroots work has shown this possibility. With plateauing of yields in the traditional green revolution belt, the rainfed agriculture of the eastern region holds the key to India's food security in the near future.

Yield is never the only objective in dryland agriculture. Dryfarming is an exercise in constrained optimisation of multiple objectives. These include household food security, fodder and firewood needs, minimum cash flow, use of available household labour etc. The risk of rainfall failure is the most enduring concern. Traditional cropping systems use diverse strategies like mixed-cropping and intercropping for rainfall insurance. The crop combinations and sequences are often highly complex and have come up taking into consideration minute variations in soil type, depth, crop maturity and susceptibility to rainfall fluctuations and household needs. As much as 60 crop mixes are reported to have existed in villages of peninsular India (Jodha, 1977). The number of crops mixed in one plot ranged from two to eight or even more. It is important to understand their value in terms of their contribution to stability of household income and maintenance of soil nutrient status. They are in great danger of being trampled over by mono-cropping practices once water becomes available.

We need not only viable agriculture packages but meticulously worked out land-use planning systems, which make careful use of available soil moisture through appropriate tree-crop mixes. Here, again, what works and what would fail will be known only after more detailed evaluation of such systems in the field. There are several dedicated institutes deputed to do this kind of work but their field presence is next to nil. Another crucial area of neglect is livestock. Small and marginal farmers and landless labourers constitute almost two-thirds of these livestock-keeping households in India. Ownership of livestock has a crucial drought-cushioning role for small and marginal farmers. With increasing grazing regulations and encroachments of common land, availability of fodder from agricultural land is one of the central concerns of poor households, particularly in drought years. Hence, their seemingly 'irrational' preference for crops with a low grain/fodder ratio. In spite of the vital role of livestock in a rural household, strategies of improving livestock productivity and health have not been systematically integrated as central interventions in the drylands. Of particular importance is the neglect of small ruminants (goats), which are the major part of livestock holdings of poor households in drylands.

The range of required interventions extends well beyond dryland production systems. They encompass many other related areas and simultaneous action is required at all levels. Provision of cheap, timely and easily negotiable credit is one such area. Drylands are poor in terms of their banking networks. Informal credit markets with interest rates ranging from 60 to 120% dominate. Organisation into self-help groups (SHGs) and linking them with banks is an effective means of credit delivery to poor house-holds. Crop insurance can be one of the financial products of SHGs.

On account of their low marketed surplus, many dryland areas have poorly developed commodity markets. Distress sale is the order of the day here, often to the same trader who is also the provider of timely credit. Creation of marketing networks through checking distress sales and pooling of whatever surplus the area produces, could be an income generating activity, which a federation of SHGs could take up. A crucial problem here is that of creating additional storage facilities for agricultural produce at the local level. Banks could finance such an activity, as shown by their recent experiments with commodity futures through a system of warehouse receipts. A much larger issue, needing greater consensus at the national level, is that of effective implementation of a support price policy and local procurement by state agencies of dryland crops. To sum up, addressing the challenge of dryland agriculture involves implementation of a package of several interlinked components.

- location-specific public investments in water infrastructure;
- soil enrichment and control of land degradation;
- agricultural package of locally appropriate seeds and low-cost, sustainable agricultural practices;
- strengthening livelihood options based on livestock, fisheries, agro-processing and forests;
- better support systems through credit, marketing, research and extension;
- mobilization of communities around natural resource rights;
- learning from local contexts about possibilities and limitations of different interventions. This applies to all actors, scientists or government representatives, NGO activists or a member of local communit

Facing the challenge of the drylands is no longer a matter of choice. It is an imperative if we are to meet the goal of national food security in the coming years. Even in the most optimistic scenario of further irrigation development in India, nearly 40% of national demand for food in 2020 will have to be met through increasing the productivity of rainfed dryland agriculture (Samaj Pragati Sahayog, 2006).

Table: *Projected Demand and Supply of Foodgrains in India in the Year 2020*

	(million tonnes)
Projected Food Demand in 2020	307
Average Food Production in Triennium Ending 2002	205
Gap to be met	102
Maximum Possible Contribution of Irrigated Agriculture of which	64
From Irrigated Area Expansion	38
From Increases in Productivity of Irrigated Agriculture	26
Minimum Balance required from the Rainfed Agriculture	38
Share of Rainfed Agriculture	37%

Source: Samaj Pragati Sahayog, 2006.

To maintain food security even at the current nutritional levels, 102 MT of foodgrains have to be produced additionally by 2020. Cropped area has plateaued in India since 1970. It has remained static at around the 140 million-hectare mark for the last three decades. This is no longer a source of increased output in Indian agriculture. As for irrigated agriculture, its contribution can arise from two sources: (*a*) expansion in the area under irrigation; and (*b*) yield improvements in the areas already under irrigation.

The ultimate irrigation potential has been estimated at 139 million hectares, of which 75 million hectares is from surface water and 64 million hectares from groundwater (Planning Commission, 2002). As per Land Use Statistics, India's gross irrigated area (GIA) was 75 million hectares in 2001, which left a balance of 64 million hectares yet to be exploited. Many states in northwest India have already exhausted their irrigation potential. Nearly 65% of this unutilised irrigation potential is in the eastern parts of the country, comprising the medium to high rainfall regions of West Bengal, Bihar, Jharkhand, Orissa, Chhattisgarh, eastern Uttar Pradesh and northern Andhra Pradesh. The GIA can be expected to reach a maximum of 100 million hectares by 2020. This is an increase of 1.25 million hectares per annum, comparable to what has been achieved historically between 1970 and 2000 (1.28 million hectares).

If irrigated area grows at this rate between 2000 and 2020, we would have an additional 25 million hectares under irrigation by 2020. Assuming the share of foodgrains in GIA does not fall, only an additional 16 million hectares of foodgrain would thereby come under irrigation by 2020. Now since much of this addition to irrigated area would be in eastern India, and given current low yields under irrigated conditions, even under a most hopeful scenario, irrigated yields here are unlikely to cross three tonnes per hectare by 2020. We can, therefore, expect expansion in irrigated area to contribute an additional 38 MT to the total annual output of foodgrains by the year 2020.

Another part of the additional food output could come from yield improvements in the areas already under irrigation. Out of the 75 million hectares of GIA, 46 million hectares were under irrigated food crops in 2000. Yields of irrigated agriculture in India began to plateau in the 1990s and have declined in some areas. Even if we optimistically assume that yield growth of 30 kg/ha is sustained over the next 20 years, the rise in yield by 2020 will be only 0.6 tonnes per hectare. Thus, areas already under irrigation could contribute an additional 26 MT of foodgrains to the shortfall in 2020. The total contribution of irrigated agriculture to foodgrain production from both area expansion and yield improvements put together is, therefore, likely to be around 64 MT, still leaving a shortfall of 38 MT of foodgrains in 2020. In other words, even in the best possible scenario of irrigation development, about 40% of the additional supply of foodgrains needed to match future rise in demand will have to come from the unirrigated segment of Indian agriculture, most of which is located in the dryland areas. Thus, from the point of view of maintaining food security, there is no alternative to raising productivity of the drylands.

Bibliography

Adrian, Cullis, and Pacey, Arnold,: *Rainwater Harvesting*, Intermediate Technology Publications, UK, 1986.

Andrews J.: *The Domesticated Capsicums*, University of Texas Press. 1995.

Ballabh Parikshit : *Global Warming: An International Emerging Issue*, Cyber Tech Pub, Delhi, 2009.

Bhandari, M.M.: *The Flora of the Indian Desert*. Scientific Publisher, Jodhpur, 1990.

Bhutani, R.C.: *Fruit and Vegetable Preservation*, Biotech Books, Delhi, 2003.

Chadha, K.L. : *Advances in Horticulture*, Malhotra Publishing House, Delhi, 2006.

Collier, W; Webb, R.: *Floods, Droughts and Climate Change*, University of Arizona Press, NY, 2001.

Das, Braja M.: *Advanced Soil Mechanics*, Taylor and Francis, Delhi, 2010.

Doijode S. D.: *Seed Germination in Fruits*, New Delhi, Malhotra Publishers, 1993.

Engelman, R., and P. LeRoy: *Sustaining Water Washington*, D.C.: Population Services International, New York, 1993.

Erik Nissen-Peterson: *Rainwater Catchment Systems*, UK: Intermediate Technology Publications, 1999.

Frasier, Gary, and Lloyd Myers: *Handbook of Water Harvesting*, U.S. Dept. of Agriculture, Agricultural Research Service, Washington D.C, 1983

Gould, John, and Erik Nissen-Peterson: *Rainwater Catchment Systems*, Intermediate Technology Publications, UK, 1999.

Goyal Sham S.: *Crop Production in Saline Environments: Global and Integrative Perspectives*, International Book, Delhi, 2004.

Gupta, P.K.: *A Handbook of Soil Fertilizer and Manure*, Agrobios, Delhi, 2011.

Hanan, J. J.: *Greenhouses: Advanced Technology for Protected Horticulture*. Boca Raton, Fla.: CRC Press, 1998.

Jackson, E.: *Crop Management and Soil Conservation*, Biotech Books, Delhi, 2011.

Jhonson, Charlotte: *Biology of Soil Science*, Oxford Book Company, Delhi, 2009.

Kanmony, J. Cyril: *Drinking Water Management: Problems and Prospects*, Mittal Pub, Delhi, 2010.

Kumar Supriya : *Amenity Horticulture, Biotechnology and Post-Harvest Technology*, Pointer, Delhi, 2006.

Lata Bhattacharya: *Biochemistry of Nutrition*, Discovery, Delhi, 2010.

Little, E.L.: *Common Fuelwood Crops*, Communi-Tech Assoc., Morgantown, West Virginia, 1983.

Malins A.: *Postharvest Handling of Pineapple and Mango*, Port of Spain, Trinidad and Tobago, 1992.

Mitra S.: *Postharvest Physiology and Storage of Tropical and Subtropical Fruits*, Oxon, CABI, 1997.

Morton, Julia F.: *Fruits of Warm Climates*. Wipf and Stock Publishers, 1987.

Narasaiah, M. Lakshmi: *Energy, Irrigation and Water Supply*, Discovery, Delhi, 2004.

Narwal, S.S.: *Allelopathy In Soil Sickness*, Scientific, Delhi, 2006.

Narwani, G.S.: *Community Water Management*, Rawat, Delhi, 2005.

Pacey A., and A. Cullis: *Rainwater Harvesting: The Collection of Rainfall and Runoff in Rural Areas,* London, U.K.: IT Publications, 1986.

Palanisami, K: *Groundwater Management and Policies*, Macmillan Publishers India, Delhi, 2008.

Prabakaran, G.: *Introduction to Soil and Agricultural Microbiology*, Himalaya, Delhi, 2004.

Ram P.C. and Chaturvedi G.S.: *Crop Production Under Diverse Environments*, Pointer Pub, Delhi, 2009.

Samuel W. Johnson: *Crops Feed from Air and Soil*, Reprint Pub, Delhi, 2005.

Sanghvi Sheela and Pahwa Prem S.: *Water Harvesting, Purification and Distribution Management* , Dominant, Delhi, 2001.

Sharma, Premjit: *Applied Soil Ecology*, Gene Tech Books, Delhi, 2007.

Singh D. P.: *Crop Production in Stress Environments: Genetic and Management Options*, Agrobios, Delhi, 2007.

Singh S.D.: *Water Harvesting in Desert*, Manak, Delhi, 1997.

Swarup, Ram: *Elements of the Nature and Prospectus of Soil*, Manglam Pub, Delhi, 2011.

Thapliyal, B.K.: *Democratisation of Water*, Serials Pub, Delhi, 2008.

Tyagi, I.D.: *Plant Breeding and Genetics at a Glance*, South Asian, Delhi, 2005.

Walton, William C.: *Groundwater Resource Evaluation*. McGraw Hill, New York, 1970.

Wilde, S.A.: *Forest Soils and Forest Growth*, Periodicals, Delhi, 1991.

Index

H

I

J

L

M

N

O

P

R

S

T

W

❑❑❑